The Truth about Science and Religion

The Truth about
SCIENCE & RELIGION

From the Big Bang to Neuroscience

FRASER FLEMING

WIPF & STOCK · Eugene, Oregon

THE TRUTH ABOUT SCIENCE & RELIGION
From the Big Bang to Neuroscience

Copyright © 2016 Fraser Fleming. All rights reserved. Except for brief quotations in critical publications or reviews, no part of this book may be reproduced in any manner without prior written permission from the publisher. Write: Permissions, Wipf and Stock Publishers, 199 W. 8th Avue., Suite 3, Eugene, OR 97401.

Wipf and Stock Publishers
199 W. 8th Ave., Suite 3
Eugene, OR 97401

www.wipfandstock.com

ISBN 13: 978-1-4982-2329-4
HB ISBN 13: 978-1-4982-2331-7

Cataloging-in-Publication data:

Fleming, Fraser

The truth about science and religion : from the big bang to neuroscience / Fraser Fleming.

xviii + 222 p. ; 23 cm. Includes bibliographical references and indexes.

ISBN 13: 978-1-4982-2329-4
HB ISBN 13: 978-1-4982-2331-7

1. Religion and science. I. Title.

BL240.3 F52 2016

Manufactured in the U.S.A.

All Scripture quotations, unless otherwise indicated, are taken from the Holy Bible, New International Version®, NIV®. Copyright ©1973, 1978, 1984, 2011 by Biblica, Inc.™ Used by permission of Zondervan. All rights reserved worldwide. www.zondervan.com The "NIV" and "New International Version" are trademarks registered in the United States Patent and Trademark Office by Biblica, Inc.™

Scripture quotations marked (CEV) are from the Contemporary English Version Copyright © 1991, 1992, 1995 by American Bible Society, Used by Permission.

Table of Contents

Foreword by Gary B. Ferngren ix
Acknowledgements xiii
Introduction xv

1. Is There Purpose to Life? Implications from the Big Bang 1
   The Big Bang and the Bible 2
   A Finely-Tuned Universe 4
   Cosmic Recycling 7
   Time before the Big Bang 8
   The Big Bang: Chance or Design? 9
   The Habitable Zone 12
   Complexity and Design 14
   Conclusion 15

2. The Origin of Life: Who or What Creates Life? 20
   Divine Creation 21
      Understanding Genesis 1 24
   Pre-biotic Evolution 26
      The Water of Life 27
      Prebiotic Evolution on an Early Earth 28
      Life's Building Blocks 31
   Divinely Guided Evolution? 32
      The Origin of Information 34
      Replication 37
      The Beginning of Life 38
      Living Cells 39
   Conclusion 42

3. Evolution: From Amoeba to Zebra 46
Cells and Organisms 46
The First Living Organisms 47
Extinction and Rebirth 49
Death, Suffering, and God 50
Natural Selection 53
Intelligent Design 55
Evolution and Creation 56
Chance and Determinacy 59
God in the Machine 61
Conclusion 62

4. Primates, Hominids, and Humans: What Makes People Human? 67
Primates and Humans 67
The Evolution of Hominids 70
The Emergence of Modern Man 73
Language 74
Humans 77
Human Development and Religion 78
Natural Evil and Good 80
Morality 83
Adam and Eve 85
Moral Evil 86
Conclusion 89

5. Jesus Christ: Prayers, Miracles, the Causal Joint, and the Resurrection 95
Prayer 99
Miracles 101
The Causal Joint 104
The Resurrection 108
Conclusion 111

6. A Brief History of Science: From Prehistory to Particle Science 115
Egyptian Science 116
Babylonian Science 117
Greek Science 118
Scientific Development in Roman, Islamic, and Indian Cultures 121

Christian Beliefs that Facilitated Science 123
   Nicolas Copernicus (1473–1543) 126
   Johannes Kepler (1571–1630) 129
   Galileo Galilei (1564–1642) 136
   Isaac Newton (1642–1727) 144
   Charles Darwin (1809–1882) 149
   Albert Einstein (1879–1955) 153
   Conclusion 158

7. **The Real Me: Mind, Brain, Soul, and Spiritual Experience 165**
   The Brain as a Computer 166
   Individual Experience of Mind 169
   Emergence: From Chemistry to Cognition 171
   Consciousness: Why Is There Self-Awareness? 173
   Mind-Altering Drugs 176
   Free Will 178
   Near-Death Experiences 181
   The Soul—The Real "Me" 183
   Religious, Spiritual, and Mystical Experiences 185
   Conclusion 189

8. **Where Science and Religion Meet: Is There Personal Relevance? 195**
   The Warfare Model of Science and Religion 195
   The Separate Spheres Approach to Science and Religion 198
   The Integration of Science and Religion 199
   Does Religion Make Any Difference? 200
   Is the Force Really with Us? 202
   Themes and Inferences 203
   Conclusion 204

9. **Epilogue: Does Science Influence Personal Belief? 210**

Bibliography 213
Index 219

Foreword

"Science and religion are intertwined like DNA," begins Fraser Fleming in his Introduction to this book. While that statement seems counterintuitive, it expresses a point of view that is both necessary and important for a correct understanding of the relationship of science and religion. For more than a century, science and religion have been thought to be in conflict, offering alternative and mutually exclusive accounts of the creation of the universe. Traditional religious narratives, like those in Genesis, have come to be considered primitive attempts of pre-scientific writers to account for the mysteries of the world around them by employing supernatural tales. Only with the rise of modern science, in this widely held view, did an accurate understanding of nature become possible. As science provided rational explanations for what had previously been thought to be God's handiwork, nature lost its mystery and religious explanations retreated into the world of the irrational. In fact, that view is a myth, based on what historians of science term Whiggism. The Whig interpretation views the past through the lens of the present and sees history as moving progressively toward the ideas and institutions of our later, more enlightened, age. Whiggish historians have sometimes distorted the past to affirm the values of the present by dividing historical figures and movements into the friends and enemies of progress.

Even more influential has been the conflict thesis, which has been for the past century the predominant view of the relationship of science and religion. It has wedded a triumphalist picture of modern science, which it views as a factually-based liberating and progressive force, with a dismissal of religion, which it sees as faith-based and regressive. The conflict thesis continues to be widely accepted; indeed, it has become the dominant narrative among both scientists and layman. But, as recent scholarship has demonstrated, it too is a myth. Throughout the past two millennia the relationship of science and religion has exhibited a multiplicity of approaches, reflecting both local conditions and particular historical circumstances. The relationship between religion and the sciences is neither a monolithic nor a static one. Both have changed over the centuries and they reflect the diverse circumstances of time and place. The popular view that the march of science is one of inexorable progress and that the controversies between religion and science were disputes in which (to quote Alfred North Whitehead) "religion was always wrong, and . . . science was always right" is based on a mistaken view of the history of scientific progress, which was as uneven as theological progress.[1]

Far from being in conflict, science and religion have often been allies and considered by their proponents to be complementary. Many leading scientists have been devout believers who studied nature (in the words of Johannes Kepler) "to think God's thoughts after him." Controversies between science and religion have tended to arise when long-accepted scientific theories were being challenged by new ones, as in the substitution of a heliocentric solar system for a geocentric universe in astronomy or in the adoption of evolutionary biology in place of a static view of biological development. Defenders of traditional scientific views have sometimes appealed to biblical texts for support against novel theories. Indeed biblical interpretation remains the crux of many disputes today between some (but by no means all) religious believers and those in the scientific community whose views they challenge. At most times in the history of Western civilization such disputes were minimal and the scientific enterprise enjoyed relatively harmonious relations with Christian thought.

In this volume Fraser Fleming casts his net broadly, while focusing on the creation of the universe and the descent of the human race. He begins by exploring the Big Bang and its implications for everything that follows. In tracing those implications philosophical and theological questions arise. What is time? When did it begin? Is the universe eternal or created? The result of chance or design? Is the universe teleological, finely tuned with the human race seemingly in mind? Chapter 2 discusses the origins of life on Earth and its religious implications. How does one harmonize the creation narrative of Genesis with what we know of

---

1. Ferngren, *Science and Religion*.

prebiotic evolution? Is evolution divinely guided or the result of chance? The discussion in chapter 3 focuses on the beginning of living organisms and the multitude of theological questions that raises that are not easily answered. Whence came death, suffering, and the extinction of species, for instance? In chapter 4 we come to the development of humanity and to another set of difficult questions. How did humans cultivate religious sensibilities? How did they develop a moral and a spiritual sense? How did moral evil first enter human society? How should we interpret the Genesis narrative of the fall of the human race and its influence for human history?

In the first four chapters Professor Fleming addresses scientific issues. In so doing he follows the traditional pattern of Christian theologians who have spoken of God's Two Books, nature and the Bible. The first book is that of *general revelation*. In Christian theology revelation is God's disclosing himself and his will to his creatures. In general revelation God reveals himself through nature. An appreciation of the natural world as God's creation has always been a central theme in Christian theology. But he also raises the questions that trouble many religious believers. Has science left any place for God in the modern evolutionary view of creation, especially in dealing with the origins of the human race? While providing an impressive and up-to-date summary of current scientific views, he demonstrates that natural science does not explain everything. For all its achievements, science does not provide ultimate answers to questions regarding the meaning of the universe or of life itself. And so he proceeds in chapter 5 to describe *special revelation,* the term that theologians use to speak of how God reveals himself in prayer, miracles, prophecy, and Scripture, which fill the gaps in general revelation.

Chapter 6 deals with the history of science and its interaction with religion from the Babylonians to the mid-twentieth century. Fleming's account is brief but provides the reader with the perspective that (as historians like to think) is necessary to understanding how we got to where we are today. In particular it demonstrates that the pioneers of modern science were not narrowly scientific in their approach, but were often men of faith who were deeply concerned with the religious implications of their scientific discoveries. And few fields of modern science have made more progress and require a religious perspective more than the neurosciences. Hence, in chapter 7 he examines the mind, the soul, and the spiritual and mystical experiences that are at the core of religious views of the world. In chapter 8, Fleming brings the various strands that he has so far dealt with together in a discussion of the way in which he believes science and religion provide a comprehensive understanding of the world around us, a world that contains both material and spiritual components. Finally, in an Epilogue (chapter 9), he provides a personal point of view. He writes both as a practicing research scientist in Chemistry and as a Christian believer who is widely read in the literature that addresses the intersection of sciences with theology. He draws on both his own experience in science and his reflections in his journey of faith.

By means of a simple analogy, Sir William Bragg (1862–1942), a Nobel laureate, likened the relation of science and religion "to the cooperation of the thumb and the fingers."[2] They are, he said, functionally and spatially opposite, but it is by means of their opposition that they are able to grasp a wide variety of objects. I find that analogy helpful. Science and religion are not adversaries. They do not offer alternative and competing views of nature. But they *are* different. When each fills the role that is intended for it, they enhance one another. On the other hand, when science attempts to make religious statements, or religion to make scientific statements, they impinge on one another's domain and thereby invite conflict. During the last two millennia they have far more often been in harmony than in conflict, each doing what the other could *not* do. In their fruitful opposition they have provided a comprehensive view of nature, and so enlarged the human mind and exalted the human spirit. The means by which they continue to accomplish that task is the subject of this book.

Gary B. Ferngren
Professor of History
School of History, Philosophy, and Religion
Oregon State University
Corvallis, Oregon

2. Grant, *The Life and Work of Sir William Bragg*.

Acknowledgements

ALMOST THREE DECADES AGO I began reading books on science and religion to understand how both ways of knowing might coexist. With a little understanding and, I confess, much hubris I began giving presentations, then a series of classes at churches, that ultimately led to co-teaching several classes on science, religion, and society with my friend and colleague Bruce Beaver. The notes I had collected ultimately formed the basis for the current book.

I began the book in collaboration with a dear friend and colleague, David Somers. The basic chronological structure developed over many hours of discussion on how to approach a book on science and religion in a form that would transcend an academic collection of ideas. As the early chapters were written Dave's focus on relevance helped keep the science in step with religion. I am also indebted to Dave in sharing his neuroscience expertise, which was enormously helpful in writing chapter 7 on mind, brain, and soul.

Over the course of writing this book I have been fortunate, blessed, to have had much advice, critical evaluation, and editorial help. My daughter, Catherine Fleming, who is working on her Ph.D. in English literature, has been my fiercest and finest critic and editor. Although I have referred to her as the Editorial Dark Lord of the North, I am extremely grateful for her refining many versions of the text. She deftly helped focus the drafts around common themes and ensured that each chapter had a central thesis. Her talent for developing ideas is much appreciated in crafting the sections into their final form.

The Reverend Dennett Beuttner, as a former lawyer and now an Anglican priest, has forced me to make sure my arguments are sound while helping me stay true to orthodox religious tenants. Dennett has a knack for straightening nuances and for viewing all sides of an argument that has influenced my thinking and helped correct at least some of the ideas I first put into print. Terry Morrison's intellectual mentoring in science and faith is deeply appreciated, as is his friendship and wisdom over many years. Rachel Luckenbill has taught me the vagaries of English grammar, though I still fall prey to loose commas! Bruce Johnson's keen intellectual insight and fine writing skills have helped tremendously in simplifying tricky concepts while staying true to the meaning I wanted to convey. I am particularly grateful to Gary Ferngren for his friendship and wisdom during my early intellectual development and for penning a thoughtful forward to the book. I am most appreciative of significant effort provided by several others who read drafts and provided valuable feedback; Catalina Achim, Alec Cleland, Iain Coldham, Brenton DeBoef, the CAFE group, Fr. James Okoye, and Howard van Cleave.

I have had the great fortune to teach a study abroad course on science and religion which used early drafts of the book. I am most appreciative of the students in these classes who have provided feedback and helped make the concepts relevant. Lastly, I thank my family for indulging many hours of writing, watching videos on science and religion, and visiting museums, exhibits, and religious sites. While I am deeply indebted to the many people who have donated their time to help craft the final manuscript, ultimately I assume full responsibility for errors in the printed version.

"To God be the glory, great things He hath done"

Traditional hymn, lyrics by Fanny Crosby 1875, first published in 1875 in Lowry and Doane's song collection, "Brightest and Best."

Introduction

SCIENCE AND RELIGION ARE intertwined like DNA. Science and religion provide two perspectives on reality that speak to life's most fundamental issues: purpose, meaning, and morality. "The Truth About Science and Religion" examines pressing issues at the intersection of science and religion by following the chronological unfolding of the universe. At the heart of many of these issues lies the central question of what being human means.

Science has become a powerful force that influences the way people think about religious issues. Extraordinary advances in science over the last two centuries have revolutionized physics, chemistry, and biology. More recently, evolutionary biology, genetics, and neuroscience have pushed the conventional boundaries of experiments with living systems. Several scientific discoveries have challenged historic theological positions through a greater understanding of reality on the one hand and through the development of techniques capable of manipulating the creation of living systems on the other. Addressing the religious ramifications of these scientific advances requires a clear understanding of both the main scientific ideas and the implications of these ideas for classical theology.

Each chapter begins by delving into the science fundamental to discussion between the scientific and religious ideas. In some chapters a rather brief introduction is all that is necessary whereas other chapters, such as the discussion of Big Bang cosmology, requires greater introduction. The style is to fairly evaluate the major themes as objectively as possible. Ideas from science that challenge conventional religious dogma are examined with the same level of criticism as religious implications of scientific discoveries. Although some author bias is inevitable, with the author having stated Christian convictions (see the epilogue), the intention is to provide a balanced presentation rather than presenting a compelling case for specific Christian beliefs or a scientific position.

Beginning with the Big Bang, the book examines the religious implications inherent in cosmology and evolution. Despite a widespread perception that science and religion are antagonists, history shows that science's development was often motivated by religious belief. Although religious motives are usually absent from recent scientific pursuits, the discoveries often raise valuable questions that impinge on religious belief. Does the vanishingly small chance of a Big Bang point to the absence or presence of God? Does natural selection render God redundant or is the exploration of biological forms under divine guidance? Following the evolution of modern Homo sapiens and the differences between humans and their hominoid predecessors, the book explores the religious dimension by focusing on good, evil, and morality. How these religious issues relate to science is examined through consideration of the life of Jesus Christ. Christ's life and teaching raises questions central to understand prayer, miracles, and the resurrection in light of modern science.

Historically, modern scientific discovery blossomed in Europe in Christian cultures that were undergoing tremendous religious change. Many early scientists held strong Christian convictions, viewing scientific study as a way to a true understanding of the world and an insight into God's character. Following the lives of several major scientists, Copernicus, Kepler, Galileo, Newton, Darwin, and Einstein, provides a brief history of science to show the influence of personal religious convictions, positive and negative, on scientific discovery. For Kepler, religious convictions provided the motivation for astronomical discovery, whereas deeper scientific study into biological evolution led Darwin from the priesthood to agnosticism.

New findings, particularly from physics and biology, are revealing a much stranger world than expected. The sun does not rise, man is genetically almost indistinguishable from advanced primates, and time and space are not what they seem. Advances in neuroscience reveal insight into human identity, causing a reappraisal of not only what being human means but personhood—the state of being a person with human characteristics and feelings. Understanding what or who controls the mental traffic in the brain impinges directly on fundamental issues of self-awareness, free will, and what happens at death. Science and religion are not only intertwined but provide mutually beneficial ways of knowing.

The Truth about Science and Religion provides a tour of how the world came to be and a framework for approaching existential questions. The book is intended to stimulate personal reflection more than providing an intellectual exercise, furnishing knowledge for personal reflection that in turn challenges core beliefs and provokes changes in behavior. Each chapter concludes with an overview that leads into a series of discussion questions for personal reflection or through a group dialogue of the religious or spiritual topics. The hope is that engagement with the ideas will facilitate individuals in developing a holistic religious and scientific mental framework for understanding of the world.

1. Is There Purpose to Life? Implications from the Big Bang

People long for understanding and meaning. Where did the world come from? What existed before there was a beginning? Is there a purpose to life? Does God exist? All attest to people's fascination with one of life's challenging questions: what, if anything, brought the world into existence? An intense explosion with precise timing and unimaginable force initiates a remarkable series of events that ultimately delivers earth: the blue planet, where butterflies dance between flowers and orcas breach seemingly for sheer delight. What a strange and beautiful world this is.

Two basic philosophical approaches have vied to explain the world's origin; either the universe always existed or the universe had a beginning. Each approach has both scientific and religious implications. These philosophies have influenced science, but science cannot provide philosophical or religious proofs. Science provides a powerful method for investigating and revealing reality with which philosophy must wrestle. Although science and philosophy may seem esoteric, distant, and impersonal, at the root of these approaches are core beliefs that influence, or should influence, every person's drive to live a life where actions are consistent with beliefs. Among the most significant of these questions is whether the world is designed and, if so, why? Alternatively, if the world is the result of chance, then how is purpose instantiated into each person's life?

THE BIG BANG AND THE BIBLE

Big Bang theory states that the universe began from a very dense, very hot "singularity." Elementary energetic particles called photons burst forth and spread out into the universe radiating energy. Cooling coalesced the photons into several larger atomic particles, quarks and gluons, that further coalesced into the three-quark structures: protons and neutrons. Over the following fifteen or so minutes, protons, neutrons, and electrons fused into the two most prevalent atomic species in the universe; hydrogen and helium. The entire sequence required less than an hour, indicating the remarkable ability of the universe's early beginnings for self-organization and development. Physicists describe the extreme choreography of the Big Bang as being seemingly programmed into the very fabric of the universe. Physicist Fred Hoyle famously ruminated that "a superintellect has monkeyed with physics."[1]

Many different pieces of evidence support the Big Bang theory. First, in the 1920s Edwin Hubble made the astounding observation that the galaxies were rapidly moving away from the center of the universe. If the universe is expanding then the natural conclusion is that sometime in the past the universe existed in a very compact form.

Scientists predicted that the enormous energy dissipating from the Big Bang would cause an afterglow in just the same way that a fire retains hot coals many hours after the last flames die. As sometimes happens in science, two groups simultaneously made the same discovery, Arno Penzias and Robert Wilson at Bell labs and Robert Dicke at Princeton; in this case finding the signature of the Big Bang as background microwave radiation. In a twist of fate, the scientists at Bell labs, while trying to develop better communication systems, found a constant background noise that could not be eradicated from their receivers. Inadvertently they had discovered the background radiation bathing the universe.

The rapid expansion of the Big Bang created an intense fireball with much of the radiation being emitted as light. God's first creative act in the Bible's opening chapter is the creation of light. Coincidence or correlation?

> In the beginning God created the heavens and the earth. Now the earth was formless and empty, darkness was over the surface of the deep, and the Spirit of God was hovering over the waters. And God said, "Let there be light," and there was light. God saw

1. Hoyle, "The Universe," 12.

that the light was good, and He separated the light from the darkness. God called the light "day," and the darkness he called "night." And there was evening, and there was morning—the first day.[2]

The grand opening lines of Genesis declare that God created the world, although without any explanation how. Believers try to harmonize the Big Bang with the Bible's famous description of God creating the world in seven days. Abundant scientific evidence for an old earth forces believers to revisit their interpretation that Genesis is literally describing seven twenty-four-hour periods. Some people concerned with maintaining the Bible's truthfulness have favored a close, literal reading of the text. For example, each "day" corresponds to millions of years. Others, who stress science as providing an equally truthful tool for understanding creation, see the first chapter of Genesis as having a poetic form not suited to a literal interpretation.

In fact, this is nothing new. Theologians since the third century have identified problems with a literal interpretation, such as there being an end to the first day without a sun or earth. A non-literal interpretation of "day" overcomes the otherwise problematic issue of God's work schedule. If God created light instantaneously, what did he do for the rest of the day? The focus in Genesis, it is suggested, is not *how* God made the world, but *that* God made the world as the stage for the drama of life.

In the 1920s Edwin Hubble's telescopic images demonstrated that the universe was continuously expanding. Prominent among the proponents of this idea was the Catholic priest and physicist, Georges Lemaître, who saw no problem harmonizing God and cosmological theory. Galaxies moving apart at the speed of light means that, playing the tape backward, there was a beginning from which all creation came. The space between galaxies is *stretching* with space continuing to grow, but exactly what is the universe expanding *into*? Like the question of what happened before the universe existed, this particular question is better suited to philosophical answers than scientific ones.

Harmonizing scriptures with new scientific discoveries is a continuous process. In a sense, the resilience of Genesis to reinterpretation as science advances shows either God's providence or people's stubborn belief in God. Harmonizing the truths of science and religion is ultimately only valuable if the result is a richer, purposeful, and more consistent life.

2. Gen 1:1–5

A FINELY-TUNED UNIVERSE

Whether experiencing nature's web in a pristine mountain glade or peering at the wonders of a working cell, evidence of an intricately functional universe is everywhere. The beautiful and elegant descriptions used of nature are exactly those used by cosmologists to describe the equations for the expansion of the universe. Equally surprising is that the mathematical equations that describe the universe's development are few and simple, the kind of equations whose discovery earns Nobel prizes.

Scientists commonly speak of equations having beauty despite the fact that no definition of beauty exists in science. Collectively, scientists agree on what constitutes a beautiful equation, an ingenious chemical reaction, or an elegant design because as humans, people see beauty in the world—the delicate lines in a face, intense colors of sunset, and the wonder of seeing a child being born. Scientists are as passionate as artists but operate within a discipline that strives for complete objectivity. Science is inherently focused on explanations of *how* the world works, but scientists, as *people*, are much more interested in understanding the *meaning* of the results. Einstein's conclusion to his first paper on general relativity captures this personal essence: "Scarcely anyone who fully understands this theory can escape from its magic."[3]

The universe not only has a beautiful mathematical structure but the equations and values are very finely tuned. Just four basic forces affected the first particles during the initial stages of the Big Bang: gravity, electromagnetism, the strong nuclear force, and the weak nuclear force. The balance between these forces is extremely precise in two ways: first the physical constants of the universe have very specific values, and second, the initial "boundary" conditions for the universe are tightly specified. Boundary conditions refer to the starting or developing nature of the universe, such as the delicate poise between expansion and collapse, and the fluctuations that form galaxies without forming black holes. Cosmologists like to say that the universe seems quite finely balanced between the outward energy of expansion and the inward pull of gravitation. Like shooting hoops, the force and trajectory must work together.

Fine tuning is nicely illustrated in the life of a star. Stars get their energy by burning hydrogen to form helium. When all the hydrogen is consumed, the core of the star pulls together under extreme gravity to form beryllium. Beryllium is a toxic element lacking the right bonding

---

3. Einstein, quoted in Chandrasekhar, "General Theory of Relativity," 4.

properties for most living organisms but is very efficiently converted to carbon (~100 percent), because there is just the right relationship between the electromagnetic and nuclear forces of beryllium and carbon. The energy for the conversion of beryllium into carbon is very closely matched so that if the conversion were only 4 percent higher or 0.5 percent lower, virtually no carbon would form. Carbon, once formed, can be consumed through a carbon-helium collision whose energy is similarly highly controlled; a deviation of only half a percent would lead all the carbon to be converted to oxygen. Carbon is slowly converted to oxygen, gradually enough to allow carbon to build up, but at a rate sufficient to produce oxygen for life. A series of delicately poised transformations provides a way for carbon to be produced from stars to provide "the building block for life."

If the value of the gravitational constant was slightly larger, then the stars' lifetimes would be much shorter with much less time for planets, and life, to evolve. Alternatively, weak gravity would mean that the stars could not generate enough heat to grow and explode to liberate the heavy atoms needed for life. How finely balanced is the force of gravity? Estimates for the allowable variation are in the range of 1 part in 100,000,000,000,000 (one hundred thousand billion).

Another example of fine tuning is the attractive force between two large masses. If this were just a little stronger, the force between the earth and the sun would be too strong and cause them to collapse into one body. If the force was just a little less, the world would spin off away from the sun. In either case the earth would not be properly warmed by the sun, and life would be unable to evolve. Owen Gingrinch, Harvard astronomer and historian of science, interprets this as follows: "Had the universe exploded with somewhat greater energy, it would have thinned down too fast for the formation of galaxies and stars. . . . Had the energy been somewhat less, gravity would have quickly got the upper hand and pulled the universe back together again in a premature Big Crunch. Like the Little Bear's Porridge, this universe is just right."[4]

Particle physicists at the supercollider in Switzerland recently found the elusive so-called "God particle."[5] Perhaps the most surprising thing about this discovery is that finding the God particle was not actually surprising. Theorists predicted the existence of a particle accounting for

---

4. Gingerich, *God's Universe*, 30.
5. Overbye, "Physicists Find Elusive Particle Seen as Key to Universe," A1.

the attraction between different mass units almost fifty years beforehand. What was surprising is the precise mass of the Higgs boson and the associated Higgs field. If the Higgs field was slightly stronger atoms would start to shrink and neutrons would decay leading ultimately to hydrogen as the only stable element. The ramifications of finding the Higgs boson and the Higgs field will keep particle physicist occupied for many years, but as quantum physics probes ever deeper into the structure of the atom, the fine tuning continues to be an integral part of the universe's structure.

The influence of philosophical ideals and scientific theory is ironically captured in the work of one of the leading physicists Fred Hoyle. Hoyle preferred to believe in the universe's eternal existence—a steady state universe—because he held strongly atheist beliefs. Hoyle showed that the light elements, particularly hydrogen, helium, and deuterium, could be formed from nuclear reactions early in the universe's existence. The intense temperatures permit nuclear fusion through particle collisions at high speeds to form the first elements of the periodic table, hydrogen and helium. Ironically, Hoyle's calculations showed that the exact amount of existing helium was best accommodated by Big Bang cosmology rather than his own favored theory of the universe's eternal existence. Hoyle's conclusion: "There is a coherent plan to the universe, though I don't know what it's a plan for."[6]

The mathematical form and values of the universal equations are not the only examples that cause scientists to say that the universe is finely tuned. The density of the universe is also strictly specified to a precision between 10^{-56} and 10^{-60}, the equivalent of hitting a bull's eye at a target twenty billion years light years away on the opposite side of the universe. Hoyle, an atheist, was so stunned by the coincidences that he wrote: "If this were a purely scientific question and not one that touched on the religious problem, I do not believe that any scientist who examined the evidence would fail to draw the inference that the laws of nuclear physics have been deliberately designed with regard to the consequences they produce inside the stars . . . ."[7]

Evolutionary biologist Richard Dawkins views the fine tuning of the universe differently. His book *The Blind Watchmaker* is subtitled "How the Evidence of Evolution Reveals a Universe without Design."[8] Dawkins

---

6. Einstein, quoted in *The Oxford Dictionary of Quotations*, 392.
7. Einstein, quoted in Stockwood, *Religion and the Scientists*, 54.
8. Dawkins, *The Blind Watchmaker*, cover.

rejects the idea that fine tuning is suggestive of a coherent plan, claiming that is instead how he would expect an evolving universe to be. The key issue is the interpretation of fine tuning in the universe; is this best explained as design imparted by God, or do godless naturalistic processes provide a better explanation for this seeming design?

COSMIC RECYCLING

Stars burn hydrogen and helium at their cores but eventually run out of fuel and burn out. Toward the end of a giant red star's life, the intense heat and pressure fuses hydrogen and helium to produce the heavier elements—carbon, oxygen, magnesium, silicon, iron, and sulfur—that comprise more than 96 percent of earth's mass.

Roughly three categories of heavy elements are present on earth. In the earth's core is a molten mass of iron while the surface mantle is rich in silicon and magnesium oxides. Sand is essentially a silicon oxide. Uranium, thorium, and potassium comprise the second category of essential elements, providing heat through radioactive decay deep inside the earth's core during the first few billion years of the earth's existence. The third set of essential elements are carbon, nitrogen, oxygen, hydrogen, and phosphorous, which comprise most of the elements in living organisms.

Exploding stars release the core elements as atomic dust that eventually cools and slowly aggregates. NASA scientists have captured spectacular images of star birth in which young stars form and simultaneously eject matter into space. The cycle by which stars explode and reform into new stars, creates an ever increasing proportion of heavier elements so that newer stars contain more heavier elements than old stars. Still, hydrogen and helium comprise *almost* 90 percent and 10 percent, respectively, of the "cosmic abundance" of the elements in our sun and the most common stars, with only traces of the heavier elements required for life. Cosmic particles ultimately experience a gravitational attraction and form a flat, rotating cloud known as a solar nebula. Nebulae evolve and form disks composed of gas, dust, and rocks orbiting a central sun.

New birth and rebirth of fundamental particles establishes a pattern that recurs throughout the universe. Hydrogen and helium form and fuse under intense pressure to form new elements that are redisbursed as old stars die and new galaxies form. Evolutionary theory is built on the same principle of death and new life that leads to better adapted organisms. Christians believe that Jesus is the pinnacle of this rebirth process, heralding the coming of a new person purged of the troubles of this world and set for eternity with God.

TIME BEFORE THE BIG BANG

Physicists wrestle with the concept of time, generally saying that the concept of a time before the Big Bang does not make sense. Einstein wrote that "People like us, who believe in physics, know that the distinction between past present and future is only an illusion, however stubbornly persistent."[9] There is no negative, backward-flowing time; time only flows forward from a certain point. And, strange as it may sound, according to modern physicists, before the Big Bang there was *no time*.

This idea is not new. Over 1,500 years ago St. Augustine thought about the perennial question of what existed before God created the world. Augustine's answer was that there can be no time without creation,[10] meaning that time is one of God's creations just as much as the physical universe. Because God created all things, including time, he reasoned that there was no time before creation: "For there was no 'then,' when there was no time."[11]

If God is outside time, then how does God experience time? The classical religious view is that God perceives all cosmic history at the same "time," raising interesting questions about free will. On one hand, God's perception seems to imply that he knows the outcome of every event, including all free choices, but on the other hand, if those choices are free and future events are truly open and changeable then how could even God know the outcome? However the question is resolved, this classical theological position places God outside time in a mysterious way in much the same fashion as modern physics places time outside the beginning of the cosmos.

Physicists have devised several clever theories that avoid defining the universe's precise beginning. In the bouncing universe scenario, the Big Bang causes an expansion just like a balloon being inflated. At a certain point inflation stops with gravity causing the universe to collapse.

---

9. Einstein, letter to Besso Family, quoted in Dyson, *Disturbing the Universe*, 193.
10. Jensen, *Divine Providence and Human Agency*, 39.
11. Augustine, *Confessions*.

The process parallels the way a deflating balloon full of air returns a rubbery mass. However the universe has the potential to repeat the process in an endless series of bang-crunch cycles. The universe exists eternally.

Eternal inflation describes an alternative beginning for the universe. A balloon-like universe continues to expand but with small patches at the surface that blister and rapidly expand to form a new universe bubble. Subordinate universes form at these attachment points which can, themselves, continue expanding in an eternally on-going process.

The Achilles heel of these theories is not so much the mathematics, challenging as it is, but the problem of verification. In a very real parallel to the problem of directly observing God, none of these theories can be observed directly. At the heart of cosmology is the difficulty of experimentally verifying processes of extreme size, heat, and density. Despite the Big Bang pushing the beginning of time back 13.7 billion years, the chain of explanation never ends. The question can always be asked: but what caused that? Ingrained into many cultures is the idea that something started this entire process, something that cannot be found using scientific laws. Ultimately individual belief is required to answer the question of where the world came from, either the universe always existed or something, God perhaps, brought the universe into being.

THE BIG BANG: CHANCE OR DESIGN?

The exquisite tuning of the universe and the amazing development of human life stuns scientists and has reinvigorated the search for life's grand purpose. Perhaps the universe's complex, intricate structure was encoded into the Big Bang in the same way that spectacular firework displays are encoded through a precisely orchestrated series of visual displays. Or perhaps this is just luck. Einstein captured this in his enigmatic way, saying "What I am really interested in is whether God could have made the world in a different way."[12]

Physicists measure the size of the universe in terms of light years—the distance light travels in one year, which gives an estimate of the universe's size at 13.7 billion light years. Compared to an individual person the universe appears astronomically large. But comparing a person to the size of their component atoms make people seem huge. An individual person lies roughly at the geometric mean size between the size of the

12. Einstein, translated in Holton, *The Scientific Imagination*, xii.

universe and an atom. The universe's size and age is intimately related. Small planets will agglomerate over time whereas large planets will collapse on themselves, which places restrictions on the type of planets capable of supporting life.

Two different approaches are taken to describe the universe's uniquely hospitable conditions for humanity's existence. The weak anthropic principle states that because there is only data for *this* universe, then scientists will inevitably find physical constants with values that allow for this universe's existence. Only because people exist can people reflect on ultimate origins. Even though the physical constants might potentially take an infinite number of values, only a few are possible because only those values allow for life to exist on earth. Ironically, this means that all of the planets and stars that pepper the night sky are part of an intricate system that is required for the universe to exist and for there to be life on earth. This is a paradoxical reversal of people's normal reaction to the universe's largess. Staring up at the night sky, people typically see the vast universe teeming with galaxies, realize their own comparative size and unimportance, and muse on the possibility for life on other galaxies. The weak anthropic principle reverses the lens to view the universe's amazing structure as a requirement for life to exist.

Cosmologists have long sought to explain the inherent order in the universe by assuming a series of successively more powerful underlying principles that reduce to just a few core mathematical equations; the Grand Unifying Theory—GUT. GUT deals with the fine-tuning dilemma by proposing that there is an underlying, and currently unknown, meta-law that exists and explains why the Big Bang would trigger a series of coincidences leading to this current world. GUT holds the potential for explaining many of the cosmic coincidences in terms of a simpler, fundamental theory. If successful, the GUT would provide a complete description at the physical level. Molecular interactions, forces, and particle properties would be fully understood and predictable. Despite the name, however, a successful GUT will still not provide sufficient detail to predict a person's every move and thought.

Developing a GUT is enormously complex. One promising approach that focuses on the intrinsic, minute structural details of atoms is string theory. String theory envisages very small particles held together by an attraction akin to that of a string of spaghetti. These very small strings resonate in many dimensions, giving rise to the properties of atoms, and are so small they are unobservable. Previously, the existence of these fundamental particles has been inferred from the detectable paths left by a particle's interaction with a photographic plate or an electronic detector. One of the difficulties with current string theory is that of experimental proof. One estimate posits that the equipment required for proving string theory would be at least ten light years in length. The point at which string theory leaves predictive science and becomes an exercise in mathematics or philosophy is a difficult question.

Physics has been extremely successful at illuminating the intricate physical relationships that govern the world's existence. Assuming that a GUT can be found, the existence of the few core principles might have arisen through a chance event that then led to the unfolding of the universe. Science can potentially uncover the underlying structure of the universe and maybe even the ultimate laws of nature. Two unanswered questions will remain: *why* do the laws have great structure, beauty, and elegance? And, how did the universe's structure arise?

The strong anthropic principle claims that humanity had to exist and therefore the universe had to be fine-tuned. A helpful analogy to understand the difference between the weak and strong anthropic principles is to imagine a person standing in front of a firing squad. One hundred sharpshooters all fire but as the smoke dissipates the person is alive. One interpretation, corresponding to the weak anthropic principle, is that the person was just incredibly lucky. An alternative interpretation, corresponding to the strong anthropic principle, is that the person had to survive; the marksmen's intention was to ensure that the person would live in just the same way that the fine tuning of the universe exists to allow life to develop.

A particularly ingenious way of requiring *this* universe to exist is to assume a multitude of universes. The multiverse theory views the 1 chance in 10^{60} as looking like incredibly good luck but with an infinite number of possible universes the chance becomes reasonable. If there exist an infinite number of universes then there must be a universe having exactly the character of our universe. The multiverse theory suffers from several unprovable assumptions, many which raise philosophical questions. Why are there random rather than non-random universes? Why are there an infinite number of universes? Furthermore, unlike most scientific theories, the multiverse theory is *not testable*.

THE HABITABLE ZONE

As stars die and explode they disperse their mass as the proverbial "dust of the stars." Subsequent accretion leads to concentric rings of increasingly dense particles that collide, stick, and fragment like breadcrumbs in a kitchen mixer. Over time the "feeding zone" generates particles ranging in size from dust grains to small planetesimals. Eventually these coalesce to form planets. Each feeding zone consists of a specific mixture of elements, with the lighter, more volatile elements being increasingly found further from the central star like ash driven from a campfire. Paradoxically, nitrogen, hydrogen, carbon, and oxygen are light elements that are more prevalent on Mars and Jupiter than earth but are essential to life on the blue planet. Had earth formed closer to the sun there would be even less of these essential elements, whereas further from the sun there would be no earth, only a planet drowned in water.

Fortunately, the accretion process generated a delightful habitat for intelligent life on earth—politicians notwithstanding! Remarkably, the earth continues to reap 40,000 tons of interstellar compost annually. Most interstellar debris is small, but occasionally large meteors penetrate the atmosphere and arrive on earth's surface. In all of earth's bombardment by meteors, one stands out; an impact 4.5 billion years ago with an accretion the size of Mars. The seemingly chance event was essential for several of earth's unique properties: the tilt axis of earth that's responsible for the seasons, the length of the day, the spin direction, and most importantly the formation of an exceptionally large moon.

Gauging the precise requirements needed for a habitable planet is difficult because there is only one vantage point in the universe: earth. From this biased perspective earth seems ideally—even providentially—positioned for life. Astrobiologists have coined the phrase "habitable zone" to describe the distance a planet needs to be from a central star for life to exist. Just like toasting marshmallows, the main issue is one of temperature: a planet too close to a sun will be fried, whereas one too far away will remain frozen. Overlaid on top of this requirement is the change of the star's luminosity over the extended periods of time required for complex life to develop. At the time of earth's formation, the sun is estimated to have been about 30 percent fainter than at present so that as the sun ages the habitable zone drifts further away from the sun. As a reference point, complex life on earth has arisen only during the last 10 percent of the earth's existence. Life can exist outside habitable zones in the same way that astronauts can exist on the moon, but this is not favorable for complex life to develop. A relatively wide habitable zone exists for microbes, which tolerate a much wider range of conditions than higher life forms, with an ever narrowing concentric habitable zone in moving up to plants and animals. Complex life, minimally animal life, requires a habitable zone where the distance of an Earth-like planet from the central star maintains an ocean of liquid water *and* an average global temperature less than 50 °C. Of all animal life on earth only a few extremophiles would be able to survive outside these conditions.

Maintaining an optimal temperature depends on the distance of a planet from the sun, which, for life to evolve, must coincide with habitable conditions on the planet's surface to support life. An aging sun releases more heat in a mad dash to use any available fuel, which moves the habitable zone further outward. The more massive the star, the faster the star brightens and the narrower the habitable zone. Although earth's sun is often viewed as a typical star, the earth's sun is larger than 95 percent of all known stars—anything but typical. The most common star in the Milky Way has only 10 percent of the mass of the earth's sun which requires a planet to be much closer to be in the habitable zone. At this close distance the gravitational tidal effect from the star induces synchronous rotation of the planet with the star so that the planet rotates with the same face toward the star. Just as with earth's moon, this synchronicity leads to excessive heating on one side of the planet while the other side freezes. Life is confined to a narrow band between the two zones.

In the same way that the earth lies in the habitable zone mapped out by the sun, the sun lies in a "galactic habitable zone" within the Milky Way. Earth's sun is about 25,000 light years from the center of the Milky Way and located between the spiral arms where the star density is relatively low. Closer to the galactic center, high energy sources bathe surrounding planets in ionizing radiation, gamma rays and x-rays, which destroy life, not to mention the occasional supernova. Ionizing radiation dissipates further from the galactic center as do the number of heavy elements required for life. Earth lies in the sweet spot in a spiral galaxy ideally positioned for life.

Dramatic pictures of galaxies beamed down from the Hubble telescope inspire awe and wonder. Some scientists assure us that these only confirm the wonder of humanity's chance existence. Is there a purpose to this series of coincidences?

COMPLEXITY AND DESIGN

Seeking understanding is an essentially human quest because people are naturally curious, pattern-finding beings. People see the stars as outlines of animals in a starry zoo. Children persistently question "Why?" Science uses design axioms to understand diverse types of complex systems. And seemingly random coincidences are interpreted as arising by design. A plane smashing into a building is a devastating accident. Two planes crashing into the same building establishes a pattern and unleashes questions: Why? Who? Patterns, even in the midst of tragedy, often cause people to search for deeper meaning in their lives.

In Victorian England, the world's complexity and structure was famously accredited to God when, in 1802, the Reverend William Paley published *Natural Theology, or Evidences of the Existence and Attributes of the Deity Collected from the Appearances of Nature*. Imagine walking across an English heath and finding a watch, "the inference we think is inevitable, that the watch must have had a maker."[13] In the same way, the integrated parts of nature bear witness to God's design. Paley's natural theology resonated with believers and remained part of apologetics training for ministers up until the turn of the twentieth century. Darwin's idea of descent with modification, however, provided a non-supernatural mechanism to explain design that ultimately became the demise of natural theology. Into the theological void sprang scientific creationism, an argument which explains the universe's order as stemming from God's omnipotent hand a mere few thousand years ago.

Battered by dramatic advances in science, scientific creationism had all but disappeared toward the latter part of the twentieth century. In the 1990s the basic argument that patterns in nature stem from an intelligent designer was reinvigorated. Some Americans with religious convictions were receptive to ideas that sought to show that the purpose evident in their personal lives was due to God's inherent design of creation. Prepared as a legal case by lawyer Phillip Johnson, the book *Darwin on Trial* argued that the fossil record did not provide sufficient *scientific* evidence to support biological evolution's claims. This was soon followed by biochemist Michael Behe's book *Darwin's Black Box*, which argued that gaps in the expected transitions arose because some biological entities were essentially *irreducibly complex* and could *only* have been made by an intelligent designer. Further support for the theory of "intelligent design" was

---

13. Paley, *The Works of William Paley*, IV.2.

advanced through a mathematical treatment of design that demonstrated how *specified complexity* is empirically detectable and therefore scientific.

The hope for an intelligent design research program has not materialized, but instead has alienated some scientists with Christian convictions.

Described as the most significant contribution to systematic theology this century, "Scientific Theology" offers a more moderate, and theologically oriented approach to the design in nature. In essence, biochemist-theologian Alister McGrath argues, the unreasonable effectiveness of mathematics and the regularity and intelligibility of nature do not *prove* the existence of God but do reinforce the plausibility of an already existing belief.[14]

This idea has grown into a more nuanced form of natural theology. This new form claims that belief in God provides a more complete and rationally persuasive view of nature that better fits with the lived human experience than a purely materialistic worldview. Today's natural theologians begin with the belief in God, then ask what kind of a world would be expected from a good God, and only then look for evidence in the world around them that will confirm their belief. Any other approach, such as Paley's, ultimately rests on an extra-religious assumption that builds a proof for God's existence on a material basis.

The new form of natural theology augments evidence for an already existing belief by showing how the world is consistent with the existence of a loving God. The approach locates natural theology as a subset of theology rather than as an independent, materialist line of inquiry. Natural theology does not offer *proof* that God exists, but rather, helps to reconcile some of the apparent contradictions between nature and theology (for example, see chapter 3 for a theological discussion of how death and suffering in an evolutionary progression might be compatible with divine creation).

CONCLUSION

Can science interpret the amazing fine tuning of the world to extract a purpose within the universe? Pascal, the brilliant mathematician and philosopher, turned to statistics for an answer: either God exists or he doesn't. If God exists and you believe in Him then the reward is life in

---

14. McGrath, *The Science of God*.

eternity. If God exists and you don't believe in Him then you end up in eternal damnation. If God doesn't exist and you believe, you've probably made a few sacrifices that you wouldn't have otherwise have made whereas if God doesn't exist and you don't believe, then you're even. If you gamble the best choice is to believe God does exist. This, however, while rational, is hardly the way that most people decide to believe in God.

Others seek purpose in the essence of nature. If people are ultimately only the product of nature and if individuals have purpose, then purpose must arise from natural processes. Does this type of purpose exist in cells or just in higher organisms? Answers to this question remain elusive.

Space agencies spend billions to find life in the outer reaches of the cosmos. If life can be found on other planets, on a remote moon, or tucked away in the corner of the galaxy then maybe the origin of life is not as special as it currently appears. The origin of life, and humanity in particular, might really be chance and a good chance at that.

What is life all about? Dramatic advances in cosmology reveal intricate details of the Big Bang. Does the fine structure of the universe impose meaning? People's answers constitute part of a lifelong quest to discover and extract meaning from life. Some people interpret events as happenstance while others attribute positive outcomes to God's providence. Theists believe that the evidence is overwhelming, whereas atheists assert that belief prevents humanity from making the next evolutionary progression to a higher form of intelligence. Who's right? Only God knows!

DISCUSSION QUESTIONS

1. Science has been fantastically successful in unlocking the secrets of nature. Why is science so effective at being able to provide answers about how the world works? Why are people able to comprehend the universe?

2. Given the predicted end of earth's solar system in about thirty million years, what is the most imperative aim for humanity?

3. Conflicts between scientific discoveries and religious doctrine have caused much difficulty for people wanting to live as intellectually honest followers of God. Are scientific discoveries ever able to change the interpretation of religion and are religious texts ever able to influence the pursuit of science?

4. Dramatic images of galaxies, star birth, and supernovae are available on the web and in television documentaries. These presentations often emphasize the awe and wonder of these images. Is this scientific or philosophical? Do these images and the feelings they invoke make you believe that they are the result of creation by God or the result of random chance?

5. If God made the universe then why isn't there clear scientific evidence for God's existence?

6. The standard Big Bang model envisages the intricate structure of billions of galaxies forming from an extremely dense, highly energetic singularity expanding in an extremely finely tuned manner. Is this equivalent to scientific belief in a miracle?

Further reading for "Is There Purpose to Life? Implications from the Big Bang"

1. Stephen Barr, *Modern Physics and Ancient Faith*. Indiana: University of Notre Dame Press, 2003. Provides an unusual blend of cosmology and Christian reflection on the meaning of the physical events. Stephen Barr is a physicist at the University of Delaware and a Catholic who writes in a very accessible style from a Christian perspective.

2. Paul Davies, *The Mind of God: Science and the Search for Ultimate Meaning*. New York: Simon & Shuster, 1992. Paul Davies is one of the few scientists who does not subscribe to conventional religion yet is adamant that the world is teaming with purpose. This book visits answers to how the world might have arisen, why the world is understandable, and why the question of meaning is so important. The style is light and engaging with a focus on ultimate meaning rather than the underlying math and physics.

3. Karl Giberson and Donald Yerxa. *Species of Origins: America's Search for a Creation Story*. Lanham, MD: Rowman and Littlefield, 2002. An excellent middle-of-the-road history of intelligent design particularly in the last three chapters.

4. Alister McGrath, *The Science of God: An Introduction to Scientific Theology*. Grand Rapids: Eerdmans, 2004. Touted as the most influential idea in theology this book is a stripped down version of a

three-volume tome. The first three chapters on background, nature, and reality are excellent and reasonably accessible.

5. Del Ratzsch, *Nature, Design, and Science: The Status of Design in Natural Science*. New York: SUNY, 2001. Provides a philosophical defense for the clearly defined study of design. The writing is dense and technical, possibly to avoid pitfalls, but the result is to limit accessibility for non-specialists.

6. Rodney Holder, *God, the Multiverse, and Everything: Modern Cosmology and the Argument from Design*. Farnham, UK: Ashgate, 2004. Physicist-Priest Rodney Holder uses a mathematical probability analysis to probe whether fine tuning is best explained by steady state theory, multiverse theory, or divine fiat. An excellent summary of current arguments is followed by a mathematical treatment, the most intense of which is relegated to appendices. Prevalent use of analogy make this quite readable.

7. Rodney Holder, *Big Bang, Big God: A Universe Designed for Life?* Oxford: Lion Hudson, 2013. Holder explains the fine tuning in the universe and compares the Christian doctrine of creation with steady state and multiverse theories. Throughout the book he shows problems with current proposals, particularly those of Stephen Hawking. He relies significantly on the use of Bayes's probability theorem as a way to argue for the reasonableness of the Christian doctrine of creation from nothing.

8. Peter D. Ward and Donald Brownlee. *Rare Earth: Why Complex Life is Uncommon in the Universe*. New York: Springer, 2007. Two experts in geology and astronomy join forces in showing just how special earth is. From earth's position in the habitable zone to early life on earth, plate tectonics, and the solar system, the authors describe a readable structure of the world that emphasizes its uniqueness.

9. Trinh X. Thuan, *Chaos and Harmony: Perspectives on Scientific Revolutions of the Twentieth Century*. Conshohocken, PA: Templeton Foundation, 2006. Combining his expertise in astrophysics with his Buddhist beliefs, Trinh draws out the world's beauty and elegance recently discovered by science. The style is engaging and poetic, readily accessible, and captures the depth and meaning in the human experience of interacting with the world.

10. Christopher Southgate, ed., *God, Humanity, and the Cosmos: A Textbook in Science and Religion*. New York: Trinity, 1999. A definitive work that is an excellent resource designed for reference rather than a continuous read.

11. Ian Barbour, *When Science Meets Religion*. New York: HarperCollins, 2000. Ian Barbour, one of the leaders in science and religion, provides an accessible book summarizing the key ideas in the field that is suited for academically inclined readers. The focus lies in showing the progression of science and religion from conflict, through independence, to dialogue, and now to integration.

12. Christopher Baglow, *Faith, Science, and Reason: Theology on the Cutting Edge*. Chicago: Midwest Theological Forum, 2009. Baglow deftly focuses on the philosophical issues emanating from the intersection of science and religion. He writes from a Catholic perspective and quotes liberally from church figures with an emphasis on Catholic writers.

2. The Origin of Life: Who or What Creates Life?

"SCIENCE WITHOUT RELIGION IS lame, religion without science is blind" wrote Einstein.[1] Pope John-Paul II refocuses Einstein's idea to show how together the two disciplines work to uncover truth: "Science can purify religion from error and superstition; religion can purify science from idolatry and false absolutes."[2] Nowhere is the intersection of science and religion more divisive than the origin of life and yet this area is where insight is most needed to guide thinking through knotty issues of genetic engineering, cloning, and stem-cell research.

Evolution is probably the greatest source of antagonism between science and religion. For religious people, God made all things. In contrast, biological evolution provides an account of life's development from inorganic matter without the necessity for any external agent. Evidence from many scientific fields, biology, geology, anthropology, paleontology, and chemistry, provides a highly plausible evolutionary sequence from Big Bang to man. Evolution is not yet supported by seamless evidence from amoeba to zebra, as there are several very significant points awaiting evidence. Nevertheless, scientific advances have been very effective in filling in many details, raising the issue of where God's influence might be.

An alternative to the explanations of a divinely created young earth or naturalistic biological evolution is an evolutionary process directed by God. Various forms of guided evolution have been proposed, ranging

---

1. Einstein, "Religion and Science."
2. John Paul II, *Letter of His Holiness John Paul II to Reverend George V. Coyne.*

from direct intervention at strategic points, to God being only the initiator of the universe's evolution. Evaluating the competing theories of earth's evolution requires objectively examining the fundamental claims of each.

DIVINE CREATION

The opening lines of the Bible set the stage for Christianity's claim that the Bible's purpose is to reveal God's love and desire for all people to live in relationship with him. Sometimes called a hymn, Genesis 1 appears to be a unique blend of prose and poetry. As poetry, the passage uses figurative language to describe God's activity by using human counterparts: speaking and seeing, working and resting. In reading the first chapter of Genesis, the question to consider is whether a poetic description of the universe's beginning could provide an accurate description of God's actions.

Genesis 1: The opening lines of the most published book in the world's history.

> In the beginning God created the heavens and the earth. Now the earth was formless and empty, darkness was over the surface of the deep, and the Spirit of God was hovering over the waters. *And God said*, "Let there be light," *and there was light. God saw that the light was good*, and He separated the light from the darkness. God called the light "day," and the darkness he called "night." *And there was evening, and there was morning—the first day.*
>
> *And God said*, "Let there be an expanse between the waters to separate water from water." So God made the expanse and separated the water under the expanse from the water above it. *And it was so*. God called the expanse "sky." *And there was evening, and there was morning—the second day.*
>
> *And God said*, "Let the water under the sky be gathered to one place, and let dry ground appear." *And it was so*. God called the dry ground "land," and the gathered waters he called "seas." *And God saw that it was good.*
>
> *Then God said*, "Let the land produce vegetation: seed-bearing plants and trees on the land that bear fruit with seed in it, according to their various kinds." *And it was so*. The land produced vegetation: plants bearing seed according to their kinds and trees bearing fruit with seed in it according to their kinds.

And God saw that it was good. And there was evening, and there was morning—the third day.

And God said, "Let there be lights in the expanse of the sky to separate the day from the night, and let them serve as signs to mark seasons and days and years, and let them be lights in the expanse of the sky to give light on the earth." *And it was so.* God made two great lights-the greater light to govern the day and the lesser light to govern the night. He also made the stars. God set them in the expanse of the sky to give light on the earth, to govern the day and the night, and to separate light from darkness. *And God saw that it was good. And there was evening, and there was morning—the fourth day.*

And God said, "Let the water teem with living creatures, and let birds fly above the earth across the expanse of the sky." So God created the great creatures of the sea and every living and moving thing with which the water teems, according to their kinds, and every winged bird according to its kind. *And God saw that it was good.* God blessed them and said, "Be fruitful and increase in number and fill the water in the seas, and let the birds increase on the earth." *And there was evening, and there was morning—the fifth day.*

And God said, "Let the land produce living creatures according to their kinds: livestock, creatures that move along the ground, and wild animals, each according to its kind." *And it was so.* God made the wild animals according to their kinds, the livestock according to their kinds, and all the creatures that move along the ground according to their kinds. *And God saw that it was good.*

Then God said, "Let us make man in our image, in our likeness, and let them rule over the fish of the sea and the birds of the air, over the livestock, over all the earth, and over all the creatures that move along the ground." *So God created* man in his own image, in the image of God he created him; male and female he created them. God blessed them and said to them, "Be fruitful and increase in number; fill the earth and subdue it. Rule over the fish of the sea and the birds of the air and over every living creature that moves on the ground." Then God said, "I give you every seed-bearing plant on the face of the whole earth and every tree that has fruit with seed in it. They will be yours for food. And to all the beasts of the earth and all the birds of the air and all the creatures that move on the ground—everything that has the breath of life in it—I give every green plant for food." *And it was so. God saw all that he had made, and it was very good. And there was evening, and there was morning—the sixth day.*[3]

The opening line of Genesis is unique among creation stories. In this and only this story God brings the universe into existence seemingly out of nothing. God's actions and the world's response, emphasized in the quotation with different type, demarcate an underlying pattern. The clear declaration that the God of the early Hebrews has made *all* of creation stands apart from the pagan myths of the neighboring prehistoric cultures. This statement of God's creative activity has always been understood as "out of nothing," a creation of matter and energy and time itself. Unlike the pagan gods who worked with pre-existing materials, God spoke and creation occurred.

The repeated phrase "*And God said*" appears at the beginning of each creative event and is followed by creation's obedience: "*And it was so.*" Capping these creative events is the declaration: "*And God saw that it was good.*" Although the sections vary in length and minor details, they follow the same pattern to reiterate that God created everything and made all things well.

The poetic structure of Genesis has been recognized for at least two millennia. As with much poetry, it has a repetitive form at several levels. On the first day God makes light and three days later he makes the heavenly lights, the sun and moon. On the second day God makes the sky and sea and three days later, the birds and fish to populate those realms. On the third day, God makes land and vegetation, the prerequisites for the land creatures and people that appear three days later on day 6. The first three days parallel the second three days: light and darkness/sun and moon, waters above and below/birds and fish, land and ocean/animals and humans. In the first three days the world is formed, while in the following three days the world is filled. The point of this unraveling symmetry is order. Each part of creation is linked together in a beautiful plan in which creative acts bring forth ecological diversity in an integrated, interdependent structure. Day seven is God's crowning glory consistent with the veneration of many ancient cultures for the number 7.

---

3. Gen 1:1–31

Understanding Genesis 1

Numerous controversies have arisen from interpretations of Genesis. The two most prevalent stem from a literal reading of the poetic style: how did God make the universe and how long did this take? A literal reading of Genesis 1 as a scientific description of God's action seems inappropriate given that the message was for a people living some time between 1500 and 500 BC. The early Hebrews' knowledge of science was minimal, but they did have a keen understanding of how God expected them to live even if they did not always follow directions! As poetry, broad statements conveying order and place in creation could truthfully provide some insight into how God made the universe without describing the precise sequence or mechanism. An advantage of couching how the universe came into being through a poetic description is that the essence of God's actions are captured in a medium that can be understood thousands of years later by modern, scientifically oriented cultures and still ring true. As poetry, Genesis is unlikely to contain specifics on the mechanism by which the universe came into being, but the language may allude to how long creation required.

Genesis implies that some creative acts were instantaneous: "'Let there be light,' and there was light."[4] Some commands imply a process of unspecified duration: "'Let the water under the sky be gathered to one place, and let dry ground appear.' And it was so."[5] Other commands initiated a creative progression: "'Let the land produce living creatures according to their kinds . . . .' And it was so."[6] The creative processes appear to require different amounts of time even though each creative event is bounded as taking a "day," which differs from the typical use of "day" to refer to a twenty-four-hour period. Words such as day, evening, and morning are still used today in broad, non-precise ways. For example, in proposing to spend the day at the beach people do not plan on arriving at midnight and staying twenty-four hours.

A non-literal interpretation of the word "day" as used in Genesis overcomes analogous problems stemming from belief in a literal twenty-four-hour day, problems such as God's work schedule. For example, if God created light instantaneously, what did he do for the rest of the day? Each creative event ends with "*And there was evening and there was*

---

4. Gen 1:3
5. Gen 1:9
6. Gen 1:24

morning—the xth day." Understanding this phrase as closing each creative event, rather than a literal description, alleviates the problem of a first "day" before the creation of sun and moon on "day" four. A poetic reading of each creative act views each day as bounding the creative periods, some short while others perhaps requiring eons.

Throughout the first half of the Bible the word "day" (*yom*) is loosely used in a variety of ways. Usually meaning a "day" of the week, the word can also mean "time," a specific "period" or "era" or a season. A natural interpretation is to view the Genesis days as metaphors for geological ages. Each Genesis day broadly correlates with a time for each creative event whether requiring millions of years or milliseconds. Reading each of the Genesis days as periods of differing creative events overcomes difficulties with a literal interpretation while preserving the intent of the chapter; God created the world.

The description of creation in Genesis 1 ends with humanity. In Genesis 2 the focus of the story is on the first man, Adam, and his companion, Eve. There are no additional depictions of "creative days," but rather events happening to humans with time being expressed in terms of a human life. Before the creation of mankind, Genesis is told from the perspective of God. After the creation of mankind, Genesis is told from the perspective of people. A reasonable interpretation of this difference is that time is described from God's perspective during days of creation and from man's perspective after creation. Einstein demonstrated that perspective means everything when considering time. The early Hebrews lived in step with the ebb and flow of the seasons, planting and harvesting according to weather rather than a set calendar. These people were less concerned with how long God worked each day and more focused on understanding God's rule and role as ultimate creator.

Genesis 1 provides a clear declaration of God as the ultimate creator. As one of the grandest poems ever written, the meaning has been grasped by diverse cultures over thousands of years. Readers of Genesis predisposed to belief in God find the poem to ring true, squarely revealing God's role despite the old-fashioned style. For others, Genesis may seem irrelevant; the Big Bang providing a better description of the universe's beginning. The challenge in reading any creation story, biblical or scientific, lies in understanding the explanations of how and why.

PRE-BIOTIC EVOLUTION

Carl Sagan grandly paraphrases the opening lines of Genesis with an alternative creation story: "The universe is all there is, all there was, and all there ever will be."[7] While religion focuses on a creator who is "outside" the universe, science's answers are from within the universe. Science's creation stories rest on a wealth of internally consistent scientific data from a diverse array of different disciplines. From astronomy to zoology, the scientific disciplines are linked through a consistent multi-billion-year development of a primeval world into an intricate web of life. In contrast to the biblical focus on *who* creates, the scientific creation story provides details of *how* the universe was created and *how long* the process took.

Radioactive dating provides a reliable method for determining the age of the earth. The method relies on the presence of several heavy elements with "extra" neutrons within the atomic core. These neutrons are the nuclear glue holding the positively charged protons together. Sometimes the repulsion between protons in the nucleus of particularly large atoms causes the atom to split into two new elements having just slightly less overall mass than the parent atom. The mass difference is emitted as energy—the radioactive decay implicit in Einstein's famous $E=mc^2$. The main isotopes for geologically timing this process are uranium, lead, potassium, and argon. At the dawn of the earth a newly formed rock would contain uranium and lead in a specific ratio that subsequently changes because of the lead that is later produced through the radioactive decay of uranium to lead. Precisely monitoring the decay of uranium into lead over short time periods provides a rate that allows the uranium-to-lead ratio to be used to date when the rock was first formed. The process is akin to knowing how far a car goes on exactly one gallon of gas and using the mileage to estimate how far the car could go on a trip with a full tank of gas.

Radioactive dating is a method that affords the age of the earth with a remarkable degree of internal consistency. Scanning the periodic table of elements identifies many radioactive nuclei which decay at different rates. Of the thirty-four radioactive nuclei only twenty-three are found in detectable amounts in nature. This is consistent with the decay of all the short-lived nuclei since the earth's formation. A few short-lived nuclei are produced by cosmic ray bombardment in the upper atmosphere, providing a continual production by a natural process. If these latter nuclei are

---

7. Sagan, *Cosmos*, 1.

eliminated from the list of persistent nuclei, then every nucleus with a half-life of less than eighty million years is missing. The earth must therefore be at least eighty million years old for the short-lived nuclei to decay out of existence. Using the radioactive dating technique with long-lived nuclei leads to the remarkable conclusion that planet earth formed about four billion years ago.

Fossils, glaciation, and the slow process of biological change hint at an ancient earth. Some religious groups believe that the earth was made in a week and is only a few thousand years old. Trying to harmonize these two beliefs is challenging. For example, if the universe were only 10,000 to 100,000 years old then what is the origin of the light from stars appearing to be billions of light years away? Did God make all the photons from the star to the earth sometime during his week's work so that the star just seems to be billions of light years away? The scenario makes God out to be deceptive. The faithful and true character of God described in the Bible is more consistent with stars being billions of light years away from the earth.

The Water of Life

Life from non-living precursors is difficult to understand. How did life emerge from star dust and expand into every nook and cranny of earth? A series of events has been proposed to explain how the atmosphere of the early earth caused such gases as hydrogen, methane, carbon monoxide, carbon dioxide, ammonia, and nitrogen to trigger a series of condensations resulting in ever larger molecules. During the Hadean era, 3.5 to 4.5 billion years ago, the earth was pummeled by asteroids in a series of violent collisions. Each impact released energy to the earth's surface, destroying even the most basic prebiotic molecules. Although these asteroids were primarily destructive, they're speculated to have brought significant quantities of ice to the developing planet. More than any other molecule, water, perhaps brought to earth as ice, is critical for life.

Water is a marvelous substance with unique properties that make water essential for life. Water has one of the highest measures of surface tension, which allows droplets to cling to leaves and enables plants to draw water up from the roots. Water's melting, boiling, and vaporization points are all much higher than those of related substances. Cooling shrinks and heating expands most materials, which is why bridges and buildings have expansion gaps. Water is anomalous in contracting until just above the freezing point where expansion occurs. The result is that ice floats on the surface of water, a property with dramatic consequences. If ice were heavier and denser than liquid water, as most solid phases are, then the ice would collect in the deepest recesses until cooling eventually turned all the lakes and the entire ocean into a solid mass. Life would be extremely difficult in a massive ice-block. Instead ice floats and, unlike water, is a poor heat conductor. Ice creates an insulating barrier between cold air above and the water below which therefore remains liquid.

Water can absorb more heat than almost any organic compound. Heat from the sun is absorbed by the oceans and lakes, providing a vast heat reservoir which moderates changes in temperature. Much energy is required to vaporize water, which makes water an excellent coolant by evaporation. Land animals make extensive use of this for cooling by sweating. An average person running for an hour would experience a fatal temperature increase of about 10 °C if they couldn't sweat. The body's five quarts of blood, largely water with a high heat absorption, counters the temperature increase and effectively cools the body through perspiration, allowing a modest overall rise in body temperature, but without frying the brain. Water has a host of unique properties: specific heat capacity, surface tension, and thermal conductivity properties, all of which conspire to make water a prerequisite for life.

Prebiotic Evolution on an Early Earth

Sometime close to 3.5 billion years ago, the earth's surface cooled to less than 100 °C allowing water to condense into vast oceans. The oceans provided a haven for simple organic molecules that would have degraded at higher temperatures. Various forms of energy bathed the primitive earth—lightning, geothermal heat, atmospheric shock waves generated by meteoric impact, ultraviolet light from the sun, and others—driving reactions in the atmosphere and ocean to form a wide variety of simple organic molecules. Among the energy options, thunderstorms are proposed as a particularly important energy source for prebiotic chemical evolution because of the efficiency of the resulting shock waves in chemical synthesis. Shock waves surpass ultraviolet light by more than a million fold in efficiently producing amino acids, leading to the conclusion that shock waves may very well have been the principal energy source for prebiotic synthesis on the early earth.

In the upper zones of this primitive atmosphere there was no ozone layer to filter living things from lethal doses of ultraviolet light. Instead, ultraviolet light irradiated the gaseous atmosphere and formed simple organic molecules; formaldehyde, hydrogen cyanide, and ammonia among others. Conversion of these simple and sometimes toxic precursors into amino acids, the building blocks of life, seems remarkably unlikely and yet is supported by some equally remarkable experiments. The classic apparatus in the famous Miller-Urey experiment consisted of a small boiling flask containing water, a spark discharge chamber with tungsten electrodes, a condenser, and a water trap to collect the products and two or more of the following gases: methane, ethane, ammonia, nitrogen, water vapor, hydrogen, carbon monoxide, carbon dioxide, and hydrogen sulfide. Although the early earth is not thought to have had a boiling ocean, the boiling action of Miller's apparatus provided a convenient means of circulating gases past the spark discharge. Perhaps even more important is the trap, which provides an efficient method for removing the mixture of products. About 2 percent of the resulting mass was in the form of amino acids. In the history of simulating prebiotic events, electrical discharge experiments have been repeated many times and consistently found to yield amino acids, the simplest building blocks required for protein synthesis.

On primordial earth, the diverse mixture of simple compounds formed in the atmosphere may have been washed down by rain into the oceans. Here life's basic units may have accumulated along with the products of ocean reactions. Further reactions inevitably took place in this reservoir, and eventually the precursor chemicals reached the consistency of a "hot dilute soup." Innumerable smaller bodies of water provided a mechanism for thickening the soup. None other than Charles Darwin first suggested a "shallow sun-warmed pond" as a place in which concentration occurred.[8] Equally likely are lakes and shoreline lagoons, with alternate flooding and evaporation to provide a constant source of chemical ingredients and concentration to allow the molecules to come together and form larger biomolecules.

The hypothetical concentration is easily envisaged in small pools, perhaps screened from ultraviolet light by overhanging rock and situated

---

8. Darwin, quoted in Ward and Brownlee, *Rare Earth*, 67.

in a warm environment as occurs naturally in countries with geothermal activity. This environment is commonly encountered in pools around Rotorua, New Zealand, and in Yellowstone National Park, although these places are inadequate for concentrating volatile substances such as aldehydes and HCN. Further concentration could occur by the accretion of organic compounds on sinking clay particles in shallow water basins. The surface of these clays can catalyze a variety of chemical reactions and could potentially condense these precursors into ever-larger molecules such as proteins and DNA.

Prebiotic evolution is not without problems. For example, carbon makes up almost 20 percent of the body's mass and yet comprises only 0.03 percent of the earth's crust. Similarly, DNA requires phosphorous in the form of phosphate, but this is one of the rarest light elements with a concentration in the earth's crust of around 1000 ppm and about 1.5 ppb in the earth's surface water. Phosphates are key constituents of not only nucleic acids but of many cell-signaling molecules. They also act as the storehouses for cells' metabolic energy. However, phosphate readily forms insoluble complexes with several metal ions, particularly calcium, thought to be present in the early earth's oceans. Access to soluble phosphates in the primitive ocean is problematic because of the prevalence of calcium and magnesium ions that readily form insoluble phosphate salts. How did such a relatively inaccessible essential element become incorporated into DNA?

The phosphorous problem and the success of the spark-discharge experiments encapsulate a fundamental principle in origin of life experiments. There is currently no direct demonstration by which simple organic molecules form selectively and then assemble into vast biopolymers having the functions found in living systems. Remarkable experiments demonstrate the viability of generating simple organic molecules, such as the amino acids from spark discharge experiments, and are suggestive of life-conferring processes. Many molecules found in living organisms are delicate, high energy species that are created by complex molecular machines, usually enzymes, that are without parallel. How these molecules formed in the absence of cellular machinery is one of the most puzzling questions for pre-biotic evolution.

Life's Building Blocks

Ingenious experiments suggest mechanisms by which simple molecules coalesce into biomolecules. Scientists might not have created life, but the synthesis of life's precursors has been clearly demonstrated in the lab. A corresponding condensation of life's building blocks from an oceanic soup would be expected to be evident from rich seams of amino acids and DNA precursors—purines and pyrimidines—all over the earth in deep sediments of great age. No confirmation of an oceanic broth has been found.

Equally important to discovering how key building blocks formed is their rate of degradation. During the Hadean era, the energy required to form prebiotic molecules would also facilitate their degradation unless some sorting mechanism were available. Several atmospheric gases are polymerized or degraded under the conditions of early earth while others would have been quickly and irreversibly converted to organic salts in the alkaline ocean. Amino acids generated at high altitudes are estimated to require roughly three years to reach the ocean, during which they are degraded by UV radiation. In one estimate no more than 3 percent are expected to survive the passage to the ocean.

Most prebiotic molecules have a limited lifetime. One recent estimate for the four core monomeric building blocks of DNA suggests lifetimes ranging from nineteen days to twelve years. At temperatures near zero Celsius, the lifetime is extended to 17,000 years. Complex molecules tend to be fragile, leading some experts to speculate that life must have formed relatively quickly after earth cooled sufficiently. Distinguished origin-of-life researcher Leslie Orgel was overheard saying, "It would be a miracle if a strand of RNA ever appeared on the primitive earth."[9]

Organic compounds degrade ever more quickly as the structures become more complex. In essence, macromolecular DNA has a much shorter "sell by" date than the small constituent nucleotide precursors. Although temperatures near freezing would give a better chance for the accumulation of the sufficient concentrations of organic compounds in the ocean, -21°C would be ideal for chemical evolution. If the early earth were some 20°C cooler than today because of less sunlight, there would be far fewer thunderstorms on the earth because thunderstorms are generated by warm, moist air coming into contact with cold, dry air. But thunderstorms are proposed as the most efficient energy source for

9. Ross, *Hidden Treasures in the Book of Job*, 121.

generating prebiotic molecules. Origin of life research is plagued by this type of quandary; thunderstorms provide the right type of energy for condensing the basic building blocks of life, but the ideal conditions for preventing the degradation of the biopolymers, DNA, and amino acids, occurs at low temperatures where thunderstorms are extremely unlikely.

The difficulty in identifying an efficient mechanism for assembling prebiotic molecules has led some people to suggest that these molecules came from outer space. Meteorites, such as the Murchison meteorite in Australia, have been found to contain amino acids. The most predominant was the simplest amino acid glycine, comprising about 40 percent of the total amino acids in the case of the Murchison meteorite. The exact amount of amino acids arriving from meteorites is under dispute but has been estimated at 0.5 g each year during the time when life first appeared. Further complicating these estimates is the contamination of meteorites with amino acids already present in earth's environment, particularly bacterially derived amino acids present in groundwater.

Life on earth is based on proteins and DNA. Forming these polymeric units involves assembling numerous precursors in a specific sequence in order to create functional biomolecules. Currently no broadly agreed sorting mechanism exists by which the correct sequence might be achieved. A similar puzzle exists at the atomic level where the formation of amino acids and nucleotides requires new molecules to form from atoms of low prevalence in the earth's crust. Evidence from spark discharge experiments provides a tantalizing mechanism to explain the formation of amino acids and at the other end of the biological spectrum, all of life rests on proteins and DNA or RNA. The transition between these points remains a great mystery.

DIVINELY GUIDED EVOLUTION?

Most scientists regard faith as something relegated to religion and are surprised to learn that science rests on several assumptions that amount to articles of faith. Belief forms the basis of scientific advances because, in proposing any hypothesis, scientists are effectively stating a belief about the world's structure. The belief may be true or contain truth, and the refining nature of the scientific method leads to an understanding based on evidence that may be far from the original belief. Scientists operate on several axioms taken on faith:

1. *Nature is Orderly*. Nature has an underlying order shown in patterns and regularities that can be discovered. The orderly structure of nature is often thought to be self-evident; yet awareness of that order is relatively recent. Kepler (1571–1630) is often credited as identifying the underlying mathematical structure of the universe, which he believed stemmed from uncovering God's purposeful design.[10] John's Gospel opens with a statement that explains the source of the purposeful order as stemming from God's nature: "In the beginning was the Word,"[11] the *logos*, the personal force, the understandable, ordered, rational principle on which all creation rests. In light of the *logos* infusing the world with rationality, including people, the validity of this understanding reflects the two-way, rational relationship intended between God and man. Einstein, in reflecting on the intelligibility of the universe and the ability to understand much of the complexity through science wrote: "God is subtle but malicious he is not."[12] In other words, the universe may be complex and contain unexpected patterns, but those are part of an orderly fundamental structure that can be understood. The world is not capricious but is a structured universe capable of being understood.

2. *Nature is Uniform*. The forces of nature are uniform throughout space and time. What happens in one laboratory in one country is reproducible under the same conditions anywhere around the world at any time.

3. *Senses Perceive Reality*. Coupled with the underlying order of nature is the ability of the human intellect to detect patterns and understand the meaning of the information inherent in the patterns. Reliable data can be obtained from the human senses or their extensions. Scientific instruments are assumed to give consistently reliable information about the way the world is despite not being able to directly "see" the object being interrogated. No-one has actually seen an electron, though scientists all believe they exist.

    Sensing reality is critical in scientific discovery because all abstract scientific discoveries are made first in the mind and then tested. Most of Einstein's work falls in this category precisely because many of his theories were counter-intuitive.

---

10. Voelkel, *Johannes Kepler and the New Astronomy*, 86.
11. John 1:1
12. Einstein quoted in Banesh Hoffmann, *Albert Einstein: Creator and Rebel*, 146.

4. *Simplicity*. If two theories or explanations fit the data, the simpler is usually to be preferred. For example, Copernicus's solar-centric system did not provide more accurate data than that of the Ptolemaic geo-centric system—the advance, recognized by mathematicians, was a simpler calculation. Similarly, the most famous scientific equation of all time, $E=mc^2$, simply and elegantly summarizes an awesome, fundamental truth underlying the universe's structure.

The axioms on which science rests are philosophical assumptions. Scientist's faith in these assumptions leads some individuals to make statements that are actually philosophical assertions. Each episode of the television show *Cosmos* began with Carl Sagan intoning that "The Universe is all there is . . ."—clearly a belief statement.[13]

Science cannot tell us why the universe is understandable or why the patterns in nature are so easily comprehended. Scientists simply make these assumptions, consciously or unconsciously, because they are so fruitful. Why people have brains capable of understanding remarkably intricate features from quantum theory to cosmology when these intellectually demanding areas have little immediate biological survival value is perplexing. From a religious perspective, the attributes of intelligence, power, and understanding are a natural consequence of people being made in God's image.

The Origin of Information

Prebiotic evolution assumes a key role of chance, in the sense of a random occurrence, to provide the right chemicals for the transition from non-living components to the first living organism. Direct laboratory simulations of conditions on an early earth must address the problem inherent in trying to reproduce a process that apparently took millions of years. Detecting chance events with small probabilities requires a long time. A scientific approach to shorten the time requires an intentional, rational experiment to replicate the "chance" events that might have produced living organisms from non-living components. Usually, experiments with low probabilities are performed under intense conditions with greater frequency to improve the chance of a favorable outcome. Experimental design involves selecting pure chemicals that are subjected to geologically plausible conditions of energy input (heat, electric discharge) and

13. Sagan, *Cosmos: A Personal Voyage*.

environment (temperature, concentration, and pH). Successful experiments generate biologically significant molecules whereas unsuccessful experiments are refined and repeated until they are successful. This repeated give and take constitutes a necessary input of information from the experimentalist and a sorting of the output to find what is experimentally relevant.

Experimentally, the sorting is provided in the analysis of the reaction mixture. Most reactions generate a mixture of products from which one or two potential precursors are carefully identified and separated. Any intervention represents an input of information. Which products are significant? This depends on what you're *looking* for, in other words, the experimental design has a specific type of product in mind for selection. This is like going to the beach and collecting shiny shells from the morass of sand and dead sea-life left along the shoreline.

The sorting process imparts information through selecting for what is important. Evolutionary models often trace the sorting mechanism to the environment, an ecological niche in biology or crystals capable of absorbing biological molecules, for example. Complex biological environments allow information to flow between organisms, such as changes to an animal's coloring to blend into the environment. In this sense biological evolution is a natural process that distills information from the environment and captures the information in the genetic code. More difficult to understand is the generation of information from simple environmental features, such as crystals, which are regular and repetitious but have minimal information content. In the beginning of earth's development there were no obvious sources of complex information. The search is for a natural process capable of amplifying minimal information inherent in minerals into complex genetic information.

The difficulty inherent in disentangling the origin of information is illustrated in spark discharge experiments. Mixtures of amino acids are generated that differ in a very subtle spatial orientation. The spatial complexity stems from an unusual peculiarity of carbon: the orientation of the four bonds allows two molecules to be assembled together with exactly the same connectivity but different arrangements in space. Each carbon center is like a hand with projecting fingers, thumb, and forearm attachments. The carbon center can have a "left" and "right" handedness, each of which naturally interlocks only with another left or right. In biological systems, the carbons of each amino acid is comprised of only one "hand." Proteins have very specific, and usually very long, sequences all with the same geometric sequence—all "lefties" in a sense. The resulting sequence imparts very specific molecular complexity, particularly near the active site of enzymes where changing just one amino acid out of hundreds can render the enzyme inactive. Spark discharge experiments generate an equal mixture of two mirror-image amino acids that are very difficult to separate because their physical and chemical properties, like melting and boiling points, are identical. Randomly incorporating amino acids of each mirror image series from a mixture also containing natural and non-natural amino acids generated in discharge experiments is not likely to lead to a functional protein.

For the chemical synthesis of proteins, all of the amino acids must have the same handedness in a very specific order. As an analogy, if a house (protein) were to be built from an array of a hundred Lego blocks comprised of twenty different colors (amino acids) then the chance of assembling only a red house would be one in 20^{100}! The chance of randomly assembling a functional protein is roughly the same as finding one grain of sand in a desert many times the size of the Sahara.

One estimate for the probability of assembling a functional enzyme through random chance puts the odds at one chance in 10^{20}. Getting the sequence right is vital because proteins serve extremely diverse biological functions. Some proteins act as enzymes, some act as ropes that anchor bone and tendons together, while others form rubber-like soft tissue that surrounds the major arteries. Random chance seems unlikely to explain the complexity required for assembling the large, "handed," three-dimensional structures so prevalent in nature. Is this the hand of God?

Assembling a functional protein requires positioning the amino acids in a specific sequence that encodes information. Enzymes contain very specific sequences of amino acids that create three dimensional "biological machines" where the sequence codes information specific to each type of enzyme. The information cannot come from some underlying attraction between amino acids because otherwise only one amino acid sequence would result. The amino acid sequence is flexible, allowing different sequences to code for different three-dimensional structures having different functions. The information coded into a protein vastly exceeds the information content of the laws governing molecular attraction. The term "specified complexity" tries to capture the meaning within a piece of information, specified in the sense of requiring a description for a specific function and complex in being unlikely to occur through chance. In this sense quartz is complex because the molecules pack into the crystal lattice in a very specific orientation, but there is minimal specificity. DNA has a very specific nucleotide order and is complex in being unlikely to have arisen by chance; other chance nucleotide orderings are possible but would not be specified. The laws of molecular attraction lead to regular repeating structures with minimal information, as found in beautiful crystals like Pyrite or fool's gold. Crystals contain one instruction repeated millions of times whereas proteins and DNA contain many different instructions depending on the one sequence specified out of the millions of possibilities.

Information is fundamental to the nature of the universe. As the universe expands there are increasing numbers of states that can potentially be adopted and so the potential for increasing information. Intelligent agents are able to distinguish whether potential states are random or contain information encoded through abstract symbolism. The ability to perform abstract thinking is the link between information and intelligence which forms the basis for mathematics, computing, and all forms of communication.

People's experience is that information comes from intelligent beings. The "Search for Extra-Terrestrial Intelligence" (SETI) involves searching for radio signals with specific patterns that convey information. Life's existence is predicated on a vast amount of information whose source remains unknown. For some, such as SETI staff, the inability to find other sentient beings simply spurs the search on to different corners of the universe. For religious believers, God is the source of the intelligence in the universe. Each individual makes a free choice, informed by experience and logic, in deciding where the universe's information comes from.

Replication

The distinction between molecules and living organisms is the information contained in the minimum number of instructions for replication. How the first proteins formed is an unsolved issue in origin-of-life research. Proteins have the remarkable ability to assemble themselves into very specific three-dimensional shapes with exactly the right groups positioned to execute their catalytic function. A protein's shape not only defines the enzymatic active site but is also changed by interactions with other cell components. Ligand binding changes the protein sufficiently to prevent enzymatic activity and effectively functions as a molecular switch, turning an enzyme on and off. The dual functionality of proteins to manipulate the cell's atomic building blocks and to provide a feedback mechanism signaling the needs of the cell has been called the second secret of life.

The same replication and feedback loops plague the formation of the first functional RNA and DNA. Polynucleotide strands require linking together one specific geometrically complex sugar with one of four nucleotide bases. Mechanisms continue to be discovered for condensing the monomeric units into the polymers required in functional DNA, with some remarkable consequences. For example, clay particles can provide active surfaces that not only facilitate the monomer condensation but also protect nucleic acids from degradation by a variety of energy sources. Small RNA sequences chosen to be partially self-complementary can self-assemble with a high preference for one monomeric "hand," providing some encouragement for the origin of RNA sequences having the same sense of handedness.

The same information requirement reoccurs in many of the biomolecules required in cells. Proteins, DNA, fats, and a myriad of cell components are built from smaller components, requiring assembly in very specific sequences for normal cell function. If DNA is the software storing the cell's information then proteins are the hardware that perform the cell's functions. Some RNA can catalyze the formation of proteins but the amino acids are not specified during this process in the same way in which cells read DNA and select amino acids for protein synthesis. The fundamental problem lies in achieving replication; DNA codes for RNA that directs protein synthesis that, in addition to acting as the catalytic cell workers, ultimately result in the assembly of DNA.

The Beginning of Life

Exactly what separates living and non-living organisms? Defining the transition from organic molecules to life is enormously difficult because living organisms exhibit such remarkable diversity of complexity. Is a virus "alive"? Is an egg alive immediately after the sperm penetrates the cell wall, or not until the fertilized egg divides? At what point do a group of cells become a baby? Even trying to define life is fraught with difficulties.

NASA provides an encompassing definition of "life [as] a self-sustained chemical system capable of undergoing Darwinian evolution."[14]

Following science's effective method of reducing a problem to the smallest discrete component, scientists have focused on the simplest expression of life in unicellular organisms. Simple cells contain four key types of complex molecules: proteins, nucleic acids (DNA and RNA), sugars (polysaccharides), and lipids. Each of these biomolecules is a polymer with a specific cell function: proteins perform the cellular reactions, nucleic acids code for the organization and replication of the cell, sugars trigger recognition events, and lipids form the core component of cell membranes.

DNA is often grandly called "the blueprint for life." But despite the remarkable code embedded within DNA's structure, DNA does not come close to the NASA definition of life. Rather, DNA behaves like many other polymers with an overall molecular motion caused by individual vibrations of the constituent atoms. Over time individual DNA molecules will move and bind to various receptors. However, the binding and recognition stems from the attraction between specific types of atoms rather than the inherent "mind" of DNA. The search for the beginning of life must, therefore, look not to smaller atomic entities but to larger structures of which DNA occupies just one critical role among many.

DNA contains an incredible amount of information. Tightly stuffed into cells, DNA would stretch to about 6 feet if drawn out from a human cell and unwound. Virtually all living organisms use DNA for storing the biological code; the chemical composition of the double helix varies between individuals and species but is universal for life on earth. How DNA came to be on earth 3.5 billion years ago is not known, but the common sequences between very diverse organisms provide an independent witness for DNA being crucial in life's beginning.

Living Cells

Scientists have sought to find the key components of life from the earliest studies of biology and chemistry. Historically, people believed that there was a "life force" inherent in living systems that was not present in inorganic materials such as rocks and minerals. Few people believe in a vital

---

14. Luisi, *The Emergence of Life*, 21.

life force any more, but, at the same time, the essential ingredients for life to appear and reproduce have also not yet been found.

Protocells represent the link between the synthesis of macromolecules and the appearance of the first living cells. Encapsulating all of the cellular components into an organized protocell is the biological equivalent of a quantum jump in understanding. The centrality of living cells has stimulated much research on the transition from cellular components to the formation of life. No other machine is known to completely assemble itself, creating one of the most daunting challenges in biochemistry. Self-replication in artificial systems is contingent on understanding this self-assembly. Understanding the evolutionary transition to replicating cells offers the potential to understand what life really is.

Cells are the central monomeric unit on which all life is based. Cells use close to a million different components and processes allowing them to function internally, to move, to signal to and find other cells, and to coalesce into multi-cellular organisms. Cells are truly remarkable nanoscale manipulators. Viewing cells gives the impression of a factory running by remote control because cells contain an enormous number of feedback loops to ensure that the right components are present in the cell. From an evolutionary perspective, the first cells would have a much simpler number of components that were later able to add additional levels of complexity.

A lipid-based cell wall provides a robust compartment capable of recognizing and excluding foreign invaders while providing a safe passage for cell metabolites. Localized within the cell are smaller entities that provide the energy for the cell (mitochondria) to synthesize proteins (Golgi), and form a central cognitive system (nucleolus). Efforts to mimic simple cells have identified reactions that can be performed inside cell walls, although these are controlled by the inherent reactivity of the chemicals rather than by a central cognitive system. Complementing this build-it-yourself strategy is the construction of artificial cells by inserting a minimal set of enzymes, nucleic acids, and cell metabolites to bring the cell to life. Simple processing by strings of RNA is possible but a great divide exists between simple chemical reactions and a cell capable of the three defining characteristics; *metabolism, self-reproduction,* and *evolution*. At the heart of the dilemma is the paradox of life: the cell components, the membrane proteins, RNA, and DNA are all interdependent. The cell wall and membrane encapsulate these key molecules in a safe environment which in turn requires proteins, DNA, and RNA for their synthesis. How cells became self-replicating is one of the most incomprehensible processes in biology.

An enormous gulf exists between simple and artificial cells and the simplest cellular organism. Genetic experiments aimed at determining how many genes are required in the simplest cell indicate that about 250 genes are minimally required for cell function. For a simple bacterium the genome consists of around 10^6 nucleotides, representing one DNA sequence out of a possible $10^{2.4 \text{ million}}$. The chance of randomly forming the genome is vanishingly small. Once the construction of the first living cell through synthetic assembly is achieved, if the endeavor is even possible, this will only provide a shadowy, though monumental, contribution to understanding the origin of life.

Perhaps the most striking aspect of the evolution of life on the earth is that it happened so fast. Scientists have suggested that life may be almost as old as the earth with an origin that may have virtually coincided with the birth of the planet. As an example, the population of organic walled microstructures from the Swaziland System, South Africa, found in 1977, was identified as the morphological remains of primitive prokaryotes. The rocks were dated as 3.4 billion years old, relatively close to the age of earth at 4.5 billion years. Despite dramatic advances in molecular biology, there is still no agreement in how life first began. Where and how life began is one of science's great mysteries. Guesses range from life being a spectacularly successful accident to being the expected outcome of a universe primed for life.

Living cells are the most complex small systems in the universe. Specialized molecules work in concert, seamlessly conveying messages to ensure that the cell performs exactly the right function within the living organism. Most perplexing is the lack of an intelligent agent controlling the cell; life is sustained and replicated by individual organisms themselves. How the first single-celled organisms came into being is a puzzle which science has been trying to unravel. At the root of the problem is a philosophical issue: from where did life's instructed complexity come? Organisms literally have a life of their own.

Information has to come from somewhere. DNA is full of information for protein synthesis, some of which forms the machinery to make and repair DNA. Random mutation can give rise to new sequences of potential information, requiring some screening process to weed out the beneficial mutations. That screening process is again another information source. Where did all the information come from in the beginning, and how did the protein-DNA symbiosis come into being? DNA is the cell's software which delivers the message for protein synthesis on the cell's main-frame. *The origin of this information-rich system is currently unknown.* Perhaps the information was primed into the universe from the beginning, although how this was done remains highly speculative.

The classic view of life's origin is through a series of key events, each building on prior levels of complexity. Empty vesicles, like oil droplets, encapsulated primitive biomolecules whose chemistry powered the cell's energy needs. During the progressive development, the enzymatic production of DNA was adopted, improved, and became an intimate part of cellular programming. The origin of many key steps in the development of life is as yet unidentified. Some argue that science will eventually be able to discover exactly how self-replicating, complex organisms first came into being. Others argue that life is too complex to understand completely and that while dramatic advances will continue, understanding life's origin will always be elusive. Both are philosophical speculations.

CONCLUSION

The origin of life requires what currently appears to be a remarkable series of coincidences. Living organisms require a series of building blocks of ever increasing size: amino acids, proteins, nucleotides, DNA, and genes. Once in place, these key cellular components must coalesce to form single-celled organisms that subsequently diverge to produce plants, animals, and ultimately the human race. Every key biological development requires a remarkable level of complexity that increases as the precursors are incorporated within ever larger structures.

Interpreting the results of origin-of-life experiments is complicated by the practical limitations of reproducing conditions of early earth on a grand scale—no one wants a Big Bang in their back yard. Complicating the analysis of these intriguing experiments is identifying the origin of the information required to assemble complex biomolecules. Purely random events generate diversity that requires a sorting mechanism to retain the information present in the new molecular entities. At the current time the sorting mechanism is unknown and the rapid emergence of life so soon after earth became habitable remains an enigma.

Science has been remarkably successful in discovering how life works; for example, the discovery of DNA, mapping of the human genome, and cloning. Discovering the origin of life would make all prior Nobel Prize discoveries pale in comparison. However, from a philosophical perspective, there is no reason to believe that science should be able to discover the origin of life, nor, even if science discovered the origins of biological organisms, would this answer the philosophical questions this book raises. The driving force to search for answers to such difficult questions as the origin of life is largely because of science's proven ability to discover new knowledge in the past and the likelihood that future benefits will accrue regardless of whether the original question is answered.

Prebiotic experiments demonstrate a remarkable synthesis of life's building blocks despite gaps in current origin-of-life theory. For some, the experiments provide a compelling explanation for the spontaneous formation of life on earth while others believe such chance occurrences require some type of divine guidance. In the past, people have suggested specific developmental interventions, formation of the eye in particular. Time has harshly treated these God-of-the-gaps arguments. More recently, God's input has been identified more within the unfolding of life on earth: God as the grand designer who continuously acts to bring the world into being. For religious people who experience God's intervention in their lives, the question naturally arises as to why God would not similarly intervene in creation. The God of Genesis stresses the relationship of people to God as the key to understanding life. The figures of speech describing God's attributes in human terms: making and speaking, conveying God as being personal and knowable, set the stage for an intimate relationship. God's "hovering over the waters"[15] makes his presence difficult to detect, and yet is consistent with God's invitation to search for him in the world, leaving the interpretation of the evidence up to each individual.

DISCUSSION QUESTIONS

1. The creation story in Genesis has often been interpreted as a literal six-day creation with the aim of creating men and women in relationship with God. How would the relationship between people and God be different if God were to create humans by a slow evolutionary process?

---

15. Gen 1:2

2. Is the idea of God consistent with an evolutionary process based on chance, waste, and suffering?

3. When a rose is picked from a rosebush, at what point is the rose "living" and "dead"?

4. A person who has just died leaves a body that no longer has life. What is the difference between the lifeless body and the virtually identical living person who inhabited the body just a few seconds or hours earlier? Is there a difference?

5. One proposal to explain the formation of high-energy biomolecules in the absence of cellular machinery is by a "frozen accident." The idea is that an accidental occurrence generates a molecule that somehow becomes codified into the living process. What is the difference between the scientific postulate of a frozen accident and a religious assertion of divine intervention in the evolutionary process?

6. What guidelines should accompany the interpretation of lab experiments designed to mimic infrequent, long-term processes?

7. Amazing experiments are being performed to understand how life began, and with remarkable success. Should there be any limits to the forms of life that scientists can create? If so what would these be and why?

8. Richard Dawkins has written that "living organisms exist for the benefit of DNA rather than the other way round."[16] Does this statement ring true with the experience of life?

9. DNA is a remarkably complex molecule performing the intricate task of replicating cellular information. Is DNA designed? What are the indicators for or against DNA being designed?

Further reading for "The Origin of Life: Who or What Creates Life?"

1. Fazale Rana and Hugh Ross, *Origins of Life: Biblical and Evolutionary Models Face Off.* Colorado Springs: NavPress, 2004. Provides a comprehensive summary of recent advances in understanding the chemical and biological origin of life with extensive references to primary literature, reviews, and conference summaries. The

---

16. Dawkins, *The Blind Watchmaker*, 126.

material is covered from a Christian perspective, and requires an undergraduate education with familiarity in science.

2. Pier Luigi Luisi, *The Emergence of Life: From Chemical Origins to Synthetic Biology*. New York: Cambridge University Press, 2006. Covers the transition from prebiotic chemistry to synthetic biology with a clear focus on identifying the origin of life. Written for graduate students, the material is intensive but clear for those with an undergraduate degree in chemistry or biology.

3. Michael Denton, *Nature's Destiny: How the Laws of Biology Reveal Purpose in the Universe*. New York: Free Press, 1998. Denton surveys a host of biological processes that point to the universe being finely tuned for the emergence of life. The fitness of a diverse set of chemical and biological processes is surveyed in an easily understandable level, and yet the material becomes somewhat overwhelming.

4. Christian de Duve, *Vital Dust: Life as a Cosmic Imperative*. New York: Basic, 1995. Biochemist and Nobel laureate de Duve surveys the rise of biomolecules through the primordial soup to the development of modern humans. De Duve does not invoke God or chance directly but seems to concede that because life is statistically unlikely the universe still seems programmed for life.

5. Paul Davies, *The 5th Miracle. The Search for the Origin and Meaning of Life*. New York: Touchstone, 1999. Science popularizer Paul Davies engages the most perplexing questions on the origin of life in an easy-to-read style. Davies weaves possible biological theories together with theories of life on Mars, Panspermia, and other planets that build on his earlier writings on cosmology. Davies is one of the finest, fairest writers with a poetic style who keeps the mystery of life and engages a few of life's grand questions along the way.

6. Dean Overman, *A Case Against Accident and Self-Organization*. Rowman and Littlefield, 1997. Overman collects a vast array of information, largely from popular books, to argue that life is too complex to have arisen by chance. Overman presents the material in short sections comprising just a few pages and marshals the arguments like a lawyer, which he is.

3. Evolution: From Amoeba to Zebra

EVOLUTION HAS PROVEN TO be the single most important theory explaining life's development. Understanding evolution at the cellular level has fostered the fields of molecular biology and genetic engineering, which provide tools for coaxing cells to develop in previously unattainable ways. In contrast, assembly of the first cell remains a deep mystery. Unlocking the *origin* of life will not only reveal how life first arose but could potentially unleash human intervention in all living processes. Francis Crick, co-discoverer of the structure of DNA, wrote that the origin of life is "almost a miracle, so many are the conditions which would have to be satisfied to get it going."[1]

CELLS AND ORGANISMS

The possibility of spontaneously forming the first cellular life from a macroscopic coalescence of molecules is generally agreed to be vanishingly small. An event of extremely low probability is possible but, even given an extremely long time, such an event remains unlikely. Just because an event is possible does not mean the event *will* happen. Terms such as "directed chance" and "biochemical predestination" have entered the scientific literature to imply that life was guided by the inherent properties of matter. The mechanism by which cells arose from inorganic precursors is sketchy but serves as a working model in the absence of a more compelling explanation.

---

1. Crick, *Life Itself*, 3.

The origin of cells is the key to understanding the origin of life. Cells are the smallest living units that can reproduce themselves from their own internal information. Virtually all cells use the same metabolic pathways, energy storage, protein replication, and genetic information system. The neurotransmitter acetyl choline is the same in organisms as diverse as plants, protozoa, and mammals. Humans share most of the same protein families with worms, flies, and plants. All these commonalities suggest a universal ancestor cell containing sufficient information to allow subsequent divergence into the three classes of living systems found on earth.

There is a tremendous amount of information in a cell. Each cell contains between 265 and 350 genes—estimated as the minimum genome size to code for sustainable, independent life. During the development of even the simplest organisms, these genes work in concert to specify organ and body structure. Organisms are constructed as integrated organs rather than as an assembly line with individual parts fitted together to make a machine. Consequently changes in one gene often have a global impact on the entire organism. Changes that are beneficial for one organ are often less beneficial for others. Evolutionary exploration through undirected random processes requires that many organisms will form and not survive before another organism arises that has an adaptive advantage.

THE FIRST LIVING ORGANISMS

Fossil and geochemical isotope evidence indicate life began on earth around 3.7 billion years ago. The development of life so soon after the earth's formation is remarkable because the early earth was heavily bombarded by meteor showers until about 3.8 billion years ago, leaving only 100 to 200 million years for life to develop. Calculations by some scientists suggest that the earliest cyanobacteria might have formed within as little as 10 million years after the earth became habitable.

Some of the best examples of these ancient fossils are stromatolites in Shark Bay, Australia. Produced by a build-up of cyanobacteria, these slow-growing micro-organisms form mat-like colonies. Ultimately, these die and form a thin, cement-like rock layer. The cyanobacteria migrate to the surface and reestablish the colony in a recurring process that eventually leads to large rock-like formations. Stromatolites flourished in the early earth and dominate the fossil record deposits from about 600 million to 2.8 billion years. During this time, stromatolites produced oxygen even in the highly reducing environment thought to exist during earth's early development.

Significant effort has been expended to identify the first living organism. Extremophiles are touted as possibly the earliest life forms. Extremophiles are microbes that live in some of the harshest environments on earth: near undersea thermal vents; up to 3.5 km below the earth's surface in hot, pressurized environments; and in geothermal hot pools. Located in extreme environments, they seem ideally suited to the conditions of early earth and yet, buried by deep sea larval vents, or within the earth's crust, they would be in a relatively stable environment. The extreme environment ironically provides protection against other hazards such as the constant meteor bombardment during the Hadean era, volcanic eruptions, devastating ultraviolet radiation, and climatic changes. Many of these microbes live at temperatures of 80–110 °C and some even thrive at temperatures as high as 169 °C. Extremophiles have unusual biochemical modifications to stabilize the organism under extreme environments of heat, salt, acidity, and cold.

Extremophiles demonstrate that life is not limited to the habitable zone of normal temperatures and pressure. Organisms living in the boiling streams in Yellowstone National park, highly acidic waste, and inside rocks deep under the earth's surface at high temperatures and pressures represent just a few of the amazing types of extremophiles found in some of earth's otherwise inhospitable places. At the same time, while extremophiles live in harsh conditions, few can tolerate a range of environmental conditions. For an extremophile to evolve into a comparable organism in a normal environment would require changes to occur in practically every protein, RNA, and ribosome. Gene sequencing places extremophiles among the lowest and shortest branches on the evolutionary tree of life. In contrast, temperate organisms appear to reside earlier in the evolutionary development of life and in the main roots of the tree of life. The evolution of the earliest living organism from an extremophile is advocated by some, but more researchers are looking for the first common organism from a surface organism in a more hospitable environment.

Fossil and geochemical records indicate that surface organisms that grow best in moderate temperatures appeared very shortly after the earth was able to support life. Once established, the first temperate organism would have had no predator and so would reproduce freely. Replication would have been limited by a food supply, which may have driven the organism to the surface. In the transition to relatively cooler temperatures, mutant organisms might have been able to accommodate the required changes to their membranes and molecular machinery, eventually leading to the three biological domains of all life on earth; archaea, bacteria, and eukaryotes.

The fossil record and DNA mapping show remarkable parallels in support of a common ancestry, an ancestry from organisms that very quickly established themselves early in the earth's beginnings. The pathway by which the first members of each biological domain developed is obscure. DNA analyses and an accumulating fossil record provide hints as to the transition to the diversity of life both past and present.

The evolution of species rests on interpreting a fossil record that is anything but a seamless web from amoeba to zebra. The Cambrian epoch, about 600 million years ago, contains all the basic animal phyla, which suddenly appear in the fossil record like an explosion rather than as a gradual progression. Subsequent geological periods chronicle the emergence and disappearance of millions of species accompanied by enormous climatic and geographical changes. Although deleterious for some, ecological changes provided new opportunities for diversity, stimulating the creative development of new life forms.

The history of life written throughout the fossil record is one of massive species loss followed by differentiation within the surviving species. The fossil record is punctuated by the sudden emergence and sudden extinction of new species rather than a gradual evolution of species—a punctuated evolution with long periods of stasis followed by intervening periods of rapid change. Dramatic changes occur before natural selection refines the new organism, which suggests that while natural selection provides a powerful mechanism for sifting changes, other influences also operate. One potential influence that might operate in concert with natural selection is the local environment which is information-laden and provides specific conditions for organisms to adapt to.

EXTINCTION AND REBIRTH

Five particularly dramatic mass extinctions are apparent from the fossil record. The extinctions correlate with powerful ecological changes: exceptional volcanic activity, a cosmic radiation event, or the impact of an asteroid or comet. The resulting dust clouds disrupted the atmosphere causing noontime darkness, plummeting temperatures, and the eventual death of many species.

The best known mass extinction occurred some 65 million years ago at the end of the Cretaceous period. Roughly two fifths of all marine animals disappeared and an even greater proportion of land animals. High iridium levels around 65 million years ago point to a massive iridium-rich meteor impacting the coast of Mexico and churning up huge dust clouds that obscured the sun. For cold-blooded animals the effect was lethal, putting an end to the dinosaurs that roamed the earth between 230 and 65 million years ago. Whereas the dinosaur age prevented the development of large mammals, their sudden loss offered new adaptations for mammals. For small, warm-blooded mammals the loss of large predators was a God-send, allowing their eventual evolution into modern Hominids.

One of the most celebrated species transitions is the origin of animals that fly. The fossilized bird *Archaeopteryx* has a body form very similar to that of a small dinosaur, teeth in the jaws, wings, and a long reinforced bony tail covered with feathers. Dinosaur fossils having remnants of feathers provide evidence that flight originated through a coincidence of body structure and covering that was adapted for a new type of motion. The cobbling together of features to allow a completely new development is characteristic of the messy way evolution proceeds with subsequent refinements over time.

DEATH, SUFFERING, AND GOD

The young Darwin is perhaps best thought of as a nominal Christian despite his early ambition to be a priest. Instead of joining the priesthood, he left to survey unknown lands on *HMS Beagle*. Over the following years he gradually discarded his Christian beliefs, in part because he was increasingly troubled by how a loving God could allow such brutality and death in the world. For Darwin, the irreconcilable difficulty was compounded by the loss of his favorite daughter, which left him angry and bitter at God. Ultimately Darwin became an agnostic.

How can God allow the loss of life caused by massive extinctions and the suffering arising from an evolutionary development of life? Developing a coherent religious framework that addresses the suffering and the loss of life in evolutionary development involves weighty issues such as the nature of good and evil, free will, and atonement, but some insight can be gleaned from a careful analysis of the second chapter of Genesis.

People often equate the Garden of Eden with an idyllic environment in which there is neither suffering nor death. Genesis 2 portrays Eden as being harmonious but not necessarily having a distinctly different biology than exists on earth today. Hints suggest that the basic ecology is the same. In Genesis 2:20 ". . . the man gave names to all the livestock, the birds in the sky and all the wild animals." There is a distinction between livestock and wild animals. Presumably these are the same types of wild, carnivorous animals alive today which survive by preying on other animals.

If God used evolution as an integral part of creation, then death, pain, suffering, and natural disasters are a part of God's creation. The fossil and biological evidence points to recurring cycles of life and death from the first appearance of living organisms. The massive species loss that occurred 65 million years ago through an asteroid impact destroyed the dinosaurs and allowed mammals to diversify into higher animals. Some are willing to accept that the species cost in the "dinosaur project" is warranted by the benefit in evolutionary exploration that ultimately led to modern, sentient beings. Others stress that God, while permitting such processes, suffers in and with creation, sharing in the pain and loss. Although there is no definitive answer to the question of why suffering and death is part of creation, wrestling with the difficulties helps better understand good and evil.

Classically, the most persuasive argument is the "free-will defense" of evil. God creates a world capable of making itself by exploring both blind alleys and new forms. Pain, suffering, and death do not predominate in nature but are necessary consequences of a process of emerging harmony which inevitably discards old forms while developing better ones. Each step is a precarious move into new, unknown, territory. Periodic destruction of life, beauty, and order by creation is a corollary of a system capable of evolving through the exploration of all possibilities. An earth in which the constant movement of tectonic plates provides a protective mantle also allows earthquakes and volcanic eruptions to unleash tremendous energy. Dwellings subject to these forces may be severely damaged leading to loss of life, a problem exacerbated by people choosing to live in densely populated buildings rather than the sparse dwellings used by primitive humans. A creation having the potential to unfold through exploration is a messy process, as most loving processes are.

Random mutation provides a supremely elegant and efficient mechanism for achieving an open, goal directed process. A world with opportunities for love, good, and free will requires an openness in which ill can occur through chance and through the choices of others. God is neither the author of evil nor directly responsible because the choices lie within creation and with created beings. Maneuverability is given for free action by intelligent beings without divine dictation. For example, the possibility of a particular natural evil such as pain allows compassion and individual expressions of sorrow, concern, and a desire to help alleviate pain. Such a process involves a cost but brings into existence states of great value: an appreciation of beauty, the possibility of moral choices, and rational understanding, for example.

Massive species loss and evolutionary development through blind chance may seem incompatible with a loving God, but the biblical picture is more complex. At the heart of Darwinian natural selection is reproduction, an essential component of God's plan for mankind as commanded in Genesis. The raising and caring of children, or any offspring, with love and nurturing is fully compatible with both evolutionary theory and theology.

Death too is part of the natural scheme of life. While death involves an inevitable loss, nature being "red in tooth and claw," the passing of one individual often provides life for another in the integrated web of life. Many of the biblical parables use the image of a seed being transformed to provide a crop for others to harvest. The image reiterates the cost often required in life. Jesus claimed that one of the highest virtues was to lay down one's life for others and provided the example himself that the world might know the full meaning of redemption.

Christian theology understands reality as an evolving creation in which creatures have the ability to change and become themselves. Science shows that the world operates as a package deal where the freedom of organisms to evolve into intellectual thinking beings comes with the potential for malevolent events. A more competent divine being could not create a universe in which there would be no disaster or disease. Such insight is the fruit of the integration model of science and religion.

Accommodating death, pain, and suffering with the possibility of free will provides an intellectual framework for understanding good and evil but offers little comfort in the face of personal tragedy. Young lives snatched away through illness or disasters raise the question "why God?" Wisdom and patience develop in the face of such evil events through the slow and difficult exercise of understanding and experience. There is no fast-track through disturbing events, but a world having this structure provides the stage on which religion can offer answers to some of life's most difficult experiences.

NATURAL SELECTION

Natural selection is the gradual process through which biological traits become either more or less common in a population because of increased or decreased reproduction rates. Inherent within natural selection is feedback from the environment that results in refinement for subsequent generations. Through competition for limited resources, the better-adapted species are more likely to proliferate and eventually eliminate poorly adapted competitors. Darwin's theory of natural selection revolutionized biology by providing a uniquely effective, unifying explanation for the diversity of life.

The fossil record shows great stability over time followed by rapid change over relatively short periods, "punctuated evolution," in which a few entities rapidly displace earlier species. Physical separation of interbreeding populations can lead to genetic divergence between two daughter populations as mutations and genetic changes accrue. Large genetic changes more readily lead to a change in the gene pool of small populations, leading eventually to a new species. Competition with the parent generation can result in annihilation of either group depending on which trait better suits the prevailing conditions.

Natural selection leads to demonstrable changes in species. What is not demonstrated in this filtering process is a drive from simplicity to complexity, that is, complex, higher-order species and systems from simple precursors. Why, for example, did this natural process give rise to greater complexity rather than greater diversity of less developed organisms? If chance alone dictated the rearrangement of genes through genetic mutation, an increase in disorder would be expected, which would degrade the complexity of living organisms. Instead, a high level of order exists. Organisms became consistently more structured and efficient.

Living systems embody an extraordinary degree of complexity that seems difficult to explain as the result of a chance series of mutations.

And yet the evolution of life is usually compared to ascending a ladder of organization with man at the top. Chance events tend to dismantle rather than build complexity. Most random mutations are deleterious rather than beneficial, causing the organism to regress, rather than exhibit superior fitness. In humans the deleterious genomic mutation rate is estimated to be as high as three new deleterious mutations per individual. The progression of the universe toward increasing disorder is captured in the second law of thermodynamics. The more complex a system, the more prone the system is to degradation and malfunction—as attested to by anyone owning electronics! Complex systems are prone to failure because disordered states far outnumber ordered states.

Evolution very successfully exploits chance mutation and natural selection in pre-existing complex systems. What is not clear is whether evolutionary processes can create new systems through random mutation and natural selection or whether evolution builds on cues already existing in the environment. For example, in early bacterial studies examining the ability of E. coli to metabolize lactose, the structural gene that codes for the production of the enzyme galactosidase necessary to metabolize lactose was deleted. Initially, the bacteria did not grow as they could not metabolize lactose, but after a few days bacterial strains emerged that did metabolize lactose. The standard interpretation of this experiment is that the bacteria tinkered with another gene; a simple mutation was made to an existing enzyme that allowed the cleavage of the bond holding the two parts of lactose together. Where did the newly mutated gene come from? Strangely, the gene was already present in the bacteria, lying dormant without serving any function. In essence, the mutation relied on the organism's resourceful use of latent information, like rediscovering a previously read book in the bookshelf.

An additional mutation is required in the gene responsible for signaling production of the newly modified enzyme. The likelihood that random chance is responsible for both the mutation in the enzyme and the signaling has been estimated to about 1 in 10^{18}, which is so small as to require about 100,000 years to achieve. For the changes to occur in the few days the experiment ran implies that spontaneous mutations are not independent events.

INTELLIGENT DESIGN

Living systems exhibit both complexity and design. Design is recognized when the event follows a *pattern* and when there is a small *probability* of an event occurring naturally. Much of the animosity toward evolution lies in the origin of design; is design a result of evolutionary refinement or a divinely guided process? A particularly frank analysis comes from the atheist Richard Dawkins who writes in *The Blind Watchmaker* that "We are entirely accustomed to the idea that complex elegance is an indicator of premeditated, crafted design. This is probably the most powerful reason for the belief, held by the vast majority of people that ever lived, in some kind of supernatural deity...."[2]

Advocates of *intelligent* design focus on complex systems, such as an eye or the molecular motor that propels bacteria, and argue that these are irreducibly complex pieces of biological machinery whose removal leads to a complete loss of function. Take out the car battery and the car stops; remove one enzyme in the bacterial flagellum and no movement occurs. Intelligent design advocates argue that irreducible complexity is evidence of a grand designer, though the movement has worked hard to avoid the claim that the intelligent designer *must necessarily* be God.

Intelligent design, ID, was launched with Michael Behe's book *Darwin's Black Box*. The "black box" represents the smallest functional unit that Behe argues cannot come from earlier precursors through gradual adaptation.[3] Instead, an irreducibly complex component is suggested as evidence of a Grand Designer's handiwork. The response to intelligent design has been rather acrimonious and tended to alienate the very people that ID proponents hoped would consider divine guidance as a plausible explanation. The entire controversy underscores the point that people interpret the complexity differently; some see complexity as evidence of God's design while others, equally in awe, attribute complex systems to chance.

Scientists have responded to arguments for irreducible complexity with counter examples in which small, beneficial advances demonstrably improve an animal's survival. Over time, accruing numerous advances leads to a functional unit that might otherwise appear irreducibly complex. For example, parts of the molecular motor allow restricted motion in other organisms as a step on the way to a molecular motor.

2. Dawkins, *The Blind Watchmaker*, xvi.
3. Behe, *Darwin's Black Box*.

Over time, science has been incredibly fruitful at explaining complex systems of the type ID advocates have identified as requiring an intervention outside the normal bounds of science. As Dawkins says, design is most often interpreted to arise from the influence of an external agent, historically God. ID enthusiasts naturally seek evidence for God in creation because they typically operate from a theistic perspective in which God's influence is confirmed through personal experience; worship, prayer, and providence. Seeking evidence for God's handiwork through divine intervention seems to be fraught with difficulty with only scientific explanations standing the test of time. A more fruitful search for God's influence in creation seems to lie in explanations of meaning and interpretations of the significance of events.

EVOLUTION AND CREATION

Evolution is often presented as the inexorable progression of life to increasingly complex systems. Evolutionary processes explore new life forms through random mutation with beneficial accrual through natural selection. As yet no inherent principle has been identified that guarantees that evolution results in increasing complexity in the way that gravity guarantees that objects will always fall towards earth. Scientific analysis of the earth's history identifies an overall trend towards increased complexity, though whether this is because evolution is somehow weighted towards increased complexity, was just luck, or was divinely ordained remains unknown. Reflecting on this point, Darwin wrote in *On the Origin of Species*, "Probably in no one case could we precisely say why one species has been victorious over another in the great battle of life."

Of all the places in which science and religion intersect, none has caused as much dissension as the theory of evolution. Religious believers have evidence of God's existence through personal experience: divine guidance, answers to prayer, and spiritual experiences. Scientists have evidence from diverse fields that evolution provides the most comprehensive, unifying explanation for life on earth. The root difficulty lies in a perception that religion and evolution provide two different and incompatible explanations for the origin of life; either God created the world with people occupying a privileged position or life developed through a completely naturalistic process devoid of supernatural intervention.

Natural selection provides an explanation of life's development while at the same time highlighting the improbability, and unpredictability, of sentient beings coming into existence. Divine guidance of life's development through an evolutionary process, theistic evolution, is an alternative theory to explain the emergence of sentient life. How this might occur is a mystery because a purely physical weighting would be both detectable and natural in the sense of being explained within the framework of science. How an infinite, non-physical God interacts with a finite creation is a perplexing issue at the forefront of science and religion. At a minimum God's causal interaction must be different from the normal physical causality.

A relationship between God and people requires the presence of causal relations between the physical entity of each individual and a divine, intangible being. For this relationship, an interrelated cause and effect must exist between God and people and between God and the physical universe. God must be able to communicate with thoughts in the brains of conscious individuals and upon the prior processes that brought sentient beings into existence. A personal creator God requires that the laws of nature are not fully deterministic but are influenced by God's intentions.

The creation story in Genesis 1 embraces both divine creation and evolutionary processes. Verse 24 indicates that God used natural process; "Let the land produce living creatures . . ." and immediately in the following verse "God made wild animals." The Bible sees no contradiction with these two statements. Apparently God unleashed a process capable of generating creatures through an exploration of the potential inherent in the world, resulting in creatures that could reproduce themselves. God brings into being a fruitful universe in which animals have the potential to develop into creatures of even greater form.

> And God said, "Let the land produce living creatures according to their kinds: the livestock, the creatures that move along the ground, and the wild animals, each according to its kind. And it was so. God made the wild animals according to their kinds, the livestock according to their kinds, and all the creatures that move along the ground according to their kinds."[4]

The vanishingly small chance of life's origin has often been explained by believers as the providential workings of a process initiated by

---

4. Gen 1:24–25

an omnipotent God. The emergence of life from a seed is a theme woven throughout the Bible that might parallel divine guidance of natural processes to bring forth life. If God has worked to bring creation into being through purely natural processes, then definitive evidence of divine intervention is likely beyond scientific detection. The best "evidence" of divine guidance might be for a process by which random events captured and enhanced evolutionary advances over time, a process believers could view as providential. Certainly the idea of spiritual direction bringing life to fruition has a long history in Christian theology. Historically, God's omnipotence was argued to be better demonstrated in divine creation of each species in turn and only later was divine activity thought to be greater through an evolving process resulting in each separate species.

The chronological development of each species is recorded in fossils laid down over time. Plants and animals whose skeletons become compressed in sandy sediment create a book whose pages are read by sequentially dating each individual layer. Reading from the lowest layer to the topmost provides a history of how the book of life has influenced the characters over time. Although some pages appear missing, a remarkably consistent and comprehensive history emerges. Recent fossil additions fill in gaps rather than causing huge upheavals to the basic ideas of evolution. Evolutionary trees trace the development from one common ancestor with a gradual divergence through the main plant and animal classes to individual species. The power of this model lies in understanding differences between animals in different parts of the tree. Why do bats fly differently than birds and why do they "nest" in the mammalian branch of the evolutionary tree rather than with the birds? Comparison of the bat's bone structure in the wing indicates a closer similarity to flying rodents, much more like an extended gossamer paw, than the bone structure of a bird.

Genetic comparisons strengthen confidence in the veracity of evolutionary trees. Each parent provides genetic information to the offspring in the evolutionary tree. Comparing genes between related animals unravels a who-dunnit mystery identifying which animal passed what genes to whom. Following this trail, with the help of hundreds of scientist-sleuths, allows an evolutionary tree to be planted. The fact that both the gene tree and the fossil record are virtually identical only serves to confirm early speculation that animals and plants evolved over time through common ancestors.

The biological description of evolutionary development is not as neat as the equations of physics, but the explanatory laws are equally powerful. For example, keen naturalists have long known that life slows down as animals get larger. Flies live for hours or days whereas turtles can live for more than a century. Surveying a wide variety of animals shows a uniform adherence to the "negative quarter-power scaling" for the relationship of metabolism to size. The number of heartbeats an animal has are roughly the same whether the animal is large or small; large animals live longer because their heartbeats are slower. This is the type of universal law that is embedded within biology, applying to animals, plants, and even bacteria. Evolution appears to be the same type of law.

CHANCE AND DETERMINACY

Chance provides the possibility for variation in different environments whereas necessity describes options bounded by natural laws. Chance does not necessarily mean a lack of design or control; a regular die is designed to make six outcomes equally likely. Chance imbues creation with diversity. A universe devoid of chance means that each event is predictable: all outcomes are known, or can be known. If life were possible in such an environment, there would be no creatures taking risks and the skills of life would be very different. In a scripted world God would be responsible for all actions, good and bad. A world without choice lacks surprises, or unique elements of character that differ and complement those of other individuals. While a predictable world might, at times, be preferable, living under such constrictions removes precisely those individual choices that make life so enjoyable and imbue life with meaning and significance. In short, chance allows individual freedom in an evolving universe.

Chance allows the possibility of events that may or may not be causally related. The collision of an enormous asteroid with early earth provides the most likely explanation for the origin of the moon. As a consequence, the earth's rotation slowed to provide twenty-four-hour periods of night and day, promoting biological evolution by providing natural periods for activity and rest. A seemingly chance event provides a fruitful outcome because of a shuffling of potentialities.

Quantum physics identifies chance being not only a result of intersecting and unrelated causal events but an intimate feature of reality.

Radioactive decay is radically indeterminate; no amount of data would allow a more accurate prediction of when a nuclei will decay. There is an irreducible degree of openness present in the world, described in quantum terms in Heisenberg's uncertainty principle, which means that a complete description of all physical motion is just not possible.

Evolution operates with the same Heisenberg-type uncertainty to explore potential new life forms. In the absence of uncertainty, life forms would be the same, whereas too much disorder would generate novelty but lack the constancy required for an organism to become established. Earth has just the right balance of constancy and novelty.

Countless examples occur in living systems where chance occurrences secure definitive outcomes. Fish produce thousands of eggs of which only a few will grow to maturity. The low probability of maturation is offset by the high number of eggs and the survival strategy insures diversity and survival of the population. Chance is harnessed within boundaries that point to a deeper element of design, order, and purpose inherent in biological systems.

The tension between chance and determinacy is essential for life to evolve. If DNA were faithfully copied with 100 percent accuracy, then life could not evolve. A delicate balance exists between accurate DNA copying that maintains an organism's integrity and "sloppy" copying that introduces beneficial variation. Complex organisms require low error rates whereas less complex organisms can have higher error rates. An error rate of roughly less than one in 10^8 is easily accommodated by higher organisms whereas bacteria, having fewer genes, can get by with higher copying errors. Paradoxically, although the earliest organisms' genomes must have been very short to avoid catastrophic copying errors, if the genome is too short then not enough information can be stored to build the copying machinery.

Infusing creation with chance means that life can be perceived as good or bad. Mutations are essential for diversity and change through natural selection. The same evolutionary mechanism that allows cells to mutate and evolve into higher life forms, creates the potential for perfectly normal cells to become cancerous. God is often blamed for not making a perfect world without disease, decay, and death. For centuries theologians have maintained that this world is the best of all possible worlds with science too revealing that the world exists as something of a package deal. The world is intricately linked as environmental issues are dramatically illustrating. Chance operates in an open system where the good cannot be completely separated from the bad.

Has evolution been biased, or are the underlying scientific laws established in such a way as to make evolution into higher Hominids a predictable outcome? Estimates at the predictability of life vary tremendously because there is no definitive answer. Re-running the evolutionary "tape of life" is expected to lead to sentient beings but in a form different from those on earth today. A movement is developing in evolutionary biology stressing the convergence of evolution to the broad typology seen throughout earth. The evidence is complex and still debated but lies in the multiple times very similar forms and advantageous traits have emerged in separate locations and at different times. The emergence of the DNA and proteineous world only two hundred million years after the Hadean era is speculated by some to be an inevitable outcome thereby explaining the convergence of all life being based on the best possible code out of 270 million alternatives. The immensity of biological hyperspace, while appearing to have vast options to explore, may actually be much more limited. Without knowledge of the probability of such processes, God's presence or absence in the process is unknown.

Any divine coaxing of the evolutionary process has so far eluded detection. An extremely subtle divine encouragement would be commensurate with the "Spirit of God" working for a divine purpose, namely, to set the stage for the drama of human existence. If life is a cosmic accident then each person is truly alone in a universe devoid of meaning. If the fabric of the universe is scripted for life then understanding the origin of life may provide insight into the character of God.

GOD IN THE MACHINE

Does God guide the process of evolution? Chance events that occur within the context of natural processes allow numerous possibilities to be explored and realized. In this sense chance is a shuffling operation that provides the opportunity for different outcomes to be explored. Each dealing of a hand of cards holds different potential for the play. The operation of chance within the limits of natural processes, converges on a limited number—perhaps only one—of potential outcomes. In a very real sense God as the creator can be viewed as guiding evolution without direct intervention.

History has dealt harshly with those who identify a specific gap in evolution with intervention by God. Quantum physics suggests two potential ways in which God may subtly interact in the world. In both cases the intervention is beyond detection because extremely small influences at the quantum level initiate domino processes that cause changes at the physical level. Chaotic processes provide one possibility where minute divine intervention could significantly influence creation while evading detection. Chaotic events are readily apparent in the world's weather patterns where large changes from a small specific input reverberate through the system in a fashion undetectable to the observer. The butterfly effect famously states that the beating of a butterfly's wings in Brazil can trigger events that change the weather in New York.[5] The small input of information may be the scientific equivalent of the Spirit working in creation.

Another approach to finding God's influence lies in the far reaches of the quantum realm. At the quantum level, action depends on probabilities of events occurring. Radioactive decay is an example. Which atom spontaneously decays cannot be predicted despite the rate of decay being statistically predictable. God's involvement at the quantum level could trigger a series of events that might appear random if analyzed scientifically, but which cause events of deep significance. In either case, the key causal joint, the point at which God interacts with the physical realm, is too subtle to detect. Faith may allow individual discernment but with sufficient unpredictability as to be ambiguous.

CONCLUSION

Death is intrinsic to evolution and yet seems anathema to a loving God. Why God allows the pain and suffering inherent in evolution is a modern twist on the perennial question of how a good God allows evil. Although there are no logically conclusive arguments, like many issues in religion, there are pointers to support belief in God. Death is necessary for life, plants must be harvested for seed to reproduce, and animal death provides protein for other animals to live. In the process, the entire created realm remains healthy by removing less fit individuals, keeping disease in check, and maintaining a balanced ecosystem.

---

5. Polkinghorne, "So Finely Tuned a Universe of Atoms, Stars, and Quanta, and God," 16.

Nature has an inherent freedom which can impact individuals positively and negatively. Cells with the ability to mutate into organisms better adapted to their environment also have the ability to mutate into cancerous cells. The same absorption of solar energy that powers wind currents to spread rain around the globe can also generate destructive tornados. Both beneficial and harmful natural processes arise from the freedom inherent in predictable laws of nature. In much the same way, personal freedom allows choices between good and evil.

Much evolutionary advantage is promoted by ruthless selfishness, a character trait at the core of sinfulness. Keeping the knowledge of a bountiful fishing spot hidden ensures continued sustenance and a fitter individual. The same characteristics that favor individual survival can be expressed at the individual level as greed, envy, manipulation, conspiracy, and exploitation of others. One role of religion is to show that these characteristics do not have to be acted on. Individual actions are not determined by evolution. People have the option to choose not to act on the arguably less noble qualities that evolution may have bestowed. Individuals control the choice to act on these impulses or not. Any inherent instinct does not absolve an individual of the moral responsibility to act above what might be called "animal instinct."

Pain and suffering are inherent ills in the world, though not without some redemption. Pain triggers an immediate response to avoid the cause: withdrawing a hand from heat, removing a splinter, shifting weight to minimize pain in a joint. Mental and physical pain cause suffering which, when endured, can build character. Without pain and suffering, life's great lessons would not likely be learned. There would be no great heroes and no great quests against the forces of wickedness. There would be no *Count of Monte Cristo*, no *Les Miserables*, no suffering Christ, and no encouragement for people to dispense love in the face of evil.

DISCUSSION QUESTIONS

1. Evolution explores new variations through seemingly blind chance. Does the seeming lack of intentionality in evolution fit with the character of the God of the Bible?
2. Will all "irreducibly complex" systems eventually be understood in terms of an evolutionary, stepwise development?

3. In the biblical story of the Israelites settling in Canaan, God says that he will not simply drive out the inhabitants, because the land will become wild before the chosen people could take over the area. Instead a gradual conquest allows the Israelites to settle in the chosen land.[6] The evolutionary process is similar in providing an environment just for humans. Did God intend this parallel, or is this pure chance?

4. Pain provides a biological mechanism for organisms to avoid bodily damage. Would God be good or evil to impart a strong sense of pain into sentient animals?

5. Darwin had four arguments for rejecting Christianity: 1. the early Genesis chapters are a "manifestly false history," 2. God as portrayed in the Old Testament was "a revengeful tyrant," 3. science makes miracles seem "incredible," and 4. the Gospels appear unreliable. Darwin also rejected eternal damnation because "almost all my best friends will be everlastingly punished."[7] Assuming that Darwin was alive today, what arguments do you think he might make for or against belief in God?

6. In supporting evolution Darwin wrote: "Why is it more irreligious to explain the origin of man as a distinct species by descent from some lower form, through the laws of variation and natural selection, than to explain the birth of the individual through the laws of ordinary reproduction?"[8] What are some possible answers to Darwin's question and what reasons support these positions?

Further reading for "Evolution: From Amoeba to Zebra"

1. Steve Stoller, *The Symphony of Creation: Science and Faith in Harmony*. Phoenix: ACW, 2002. Using musical metaphors, Stoller addresses many of the main issues of science and religion. Chapter 7 focuses on the interplay of good, evil, and free will within nature.

2. Kenneth Miller, *Finding Darwin's God: A Scientist's Search for Common Ground between God and Evolution*. New York: Harper Perennial, 1999. Miller is a biologist and a Catholic who passionately

---

6. Deut 1–4
7. Darwin, quoted in Thomson, *Private Doubt, Public Dilemma*, 80.
8. Darwin, quoted in Phipps, *Darwin's Religious Odyssey*, 125.

argues for a scientifically credible approach to evolution for religious believers. Miller is at his best when interpreting biological experiments. The later more philosophical parts of the book provide the author's speculations on theistic evolution.

3. Fazale Rana and Hugh Ross, *Origins of Life: Biblical and Evolutionary Models Face Off*. Colorado Springs: NavPress, 2004. Provides a comprehensive summary of recent advances in understanding the chemical and biological origin of life with extensive references to primary literature, reviews, and conference summaries. The material is covered from a Christian perspective and requires an undergraduate education in science to follow many of the arguments under discussion.

4. Michael Denton, *Nature's Destiny: How the Laws of Biology Reveal Purpose in the Universe*. New York: Free, 1998. Denton surveys a host of biological processes that point to the universe being finely tuned for the emergence of life. The fitness of a diverse set of chemical and biological processes is surveyed in an easily understandable level, although the volume of material can be overwhelming.

5. Ian Tattersall, *Paleontology: A Brief History of Life*. Conshohocken, PA: Templeton Foundation, 2010. Tattersall races through the evolution of life from the earliest rocks and fossils to the arrival of man a few billion years later. Tattersall, a curator at the Museum of Natural History in New York, chronicles the rise and fall of species by focusing on the beneficial traits that accrue over the millennia. A distinct feature of the book is the amount of knowledge woven through many diverse fields.

6. Richard Swinburne, *Providence and the Problem of Evil*. Oxford: Clarendon, 1998. Developing an understanding of good and evil is one of the most difficult philosophical problems that plagues the intellectual development of each person. Swinburne argues that God wants people to freely choose good over evil, to form character, and allow love. Within this perspective, such choices are only possible when the real possibility exists for evil through bad choices.

7. Keith Ward, *God, Chance, and Necessity*. London: Oneworld, 1996. Keith Ward deftly identifies the philosophical assumptions behind the Big Bang and evolution and then takes Peter Atkins and Richard Dawkins to task for promoting philosophical ideas from unsound

logic. Ward is at his best identifying philosophical space where God may play a role in guiding evolution without being a benign dictator.

8. Simon Conway-Morris, *Life's Solution: Inevitable Humans in a Lonely Universe*. New York: Cambridge University Press, 2003. The book provides an excellent overview of evolutionary processes by confronting major weaknesses in the development of life. Conway-Morris argues that, despite difficulties, life was destined to arise and result in sentient beings.

4. Primates, Hominids, and Humans: What Makes People Human?

AN OBVIOUS RELATIONSHIP EXISTS between humans and other primates, although the nature of the relationship is not obvious. Fierce arguments have raged over whether humans are animals, naked apes, beings made in God's image, or a mixture of all three. The scientific explanation for the evolution of primates and the origin of modern civilization appears at odds with a cursory reading of Genesis. Any attempt at reconciling biology's ever-increasing discoveries with religious insight on human origins requires understanding the complexity of hominid development, wrestling with what being human means, reading Genesis to identify the main themes being conveyed in the creation of Adam and Eve, and exploring potential relationships between the Genesis account and the scientific description of hominid evolution. The following chapter spends considerable time surveying the cognitive abilities of early hominids as a prelude to understanding what being human means in religious and moral terms, and how the ideas relate to the biblical story of Adam and Eve.

PRIMATES AND HUMANS

Dramatic climate change 5–10 million years ago provided conditions for evolutionary innovation and the emergence of a diverse array of primates. Primates range from the tiny tarsier of Asia, the lemurs of Madagascar, and the bush babies of Africa to chimps and modern humans. Despite the differences, these assorted primates are linked to a common ancestor with advanced grasping and manipulation ability and a considerably enlarged brain in which vision was more important than smell.

Historical attempts to measure the cognitive ability among primates focused on skull comparison. However, skull size does not correlate with the cognition differences between chimpanzees and hominids, or even among hominids. Some early hominids had larger brain cavities but exhibited significantly less cognitive ability than the current human population. Even among earth's current population there is considerable variation; although Einstein was perhaps the greatest genius of the last century, his brain weighed only 88 percent of the average.

Cognition differences between chimps and humans are evident from comparisons of physiological development of the two species. Chimps are among the most advanced living primates and have brains weighing roughly one third the weight of *Homo sapiens*. The relative size of a chimp's frontal lobe is almost the same as that of *Homo sapiens* but a chimp brain is not just a scaled down version. Allowing for comparative differences in body size, chimpanzee brain development is rapid, progressing from 40 percent of the size of an adult chimp brain at birth to 80 percent after only a year. In contrast, a human brain is only 25 percent the size of an adult brain at birth and requires a decade before reaching 95 percent of the size of an adult brain. The relatively slow maturation, development of language skills, and mental processing are critical in each person's development into an intelligent, free-thinking, autonomous individual.

A qualitative cognitive gulf separates humans from their closest primate relatives. Chimps express basic ideas showing cause and effect—through their use of rudimentary tools—but lack the much longer chains of cause and effect that humans routinely employ to make decisions. Chimps acquire the capacity to manipulate objects and can communicate through a sign-language system taught by humans. Once chimps acquire the skills to communicate with their human trainers, they have little to say unless prompted and do not communicate through any analogous system in the wild.

Structurally the brains of primates are similar, as they must be to control many of the same biological functions; there is no "human lobe" that confers distinctly human characteristics. Scientists take two fundamentally different approaches to understanding the cognitive differences between higher primates and humans; looking ever more closely at brain structure as a source of the difference or looking toward holistic characteristics, particularly those showing a high level of retrospection. Scientists focusing on genes or brain structure can sometimes identify differences between species that may result in primate's reduced cognition relative to modern people. Social scientists complement this approach with experiments that provide insight into how primates and humans think.

Chimps perform well in some cognition experiments but not others. Physical cognition tests such as using a stick to retrieve a reward are performed equally well by chimps or people whereas social cognition tests such as learning a solution by observation are performed better by people. Extrapolate this to cultural learning and a greater gulf emerges. Chimps pass on behaviors that they might stumble across in a lifetime but the accrual of advantageous behaviors is slow and sporadic, and chimps have difficulty transferring these ideas to different communities. In contrast, people amass and manipulate an enormous collective cultural influence: language, mathematics, history, government, and an array of similar cultural products. The ability to learn and to manipulate language creates a dramatic difference between primates and modern humans which allows abstract mental concepts to be passed on to others, concepts ranging from tool production to ideas of the afterlife.

There is a minimum neuronal interconnection within the cerebral cortex required for complex mental construction. Many higher animals are capable of constructing mental scenes. Dogs that slink back to their owners may demonstrate that they understand the behavior they were supposed to exhibit and their choice of a disobedient action instead. Anecdotes of chimps hiding their emotions, covering compromising body parts with their hands, or feigning disinterest in a human observer before unleashing a mouthful of water on an unsuspecting victim, show that chimps are fully aware of how others understand their actions. Animals that practice deception show an ability to construct two mental scenes, one with the behavior others expect them to exhibit and another mental framework with the creative act of deception.

In cases where chimps, or other animals, act in ways tantamount to moral behavior, are their actions determined by the same type of deliberation and internal reflection that occurs in people? Until very recently, animals were not thought to understand social justice but several experiments suggest otherwise. Capuchin monkeys rewarded for accomplishing a task display dramatic outbursts when a cage-mate is rewarded for the same task with a food treat of comparably greater value. The monkey's action appears to show righteous indignation and a perception of justice typically associated with people rather than with animals.

Historically, self-consciousness was identified as a human characteristic at the core of what separates humans from other animals. Despite many people's desire for a clear dividing point between the way humans and animals think, experiments continue to emerge that show diverse human traits in many types of animal behavior. Early modern humans may have begun a process of internal reflection because humans are social beings who prefer to live in close communities. Sympathy, the ability to interchange mental frameworks with another, has been suggested as an early trait developed by socially oriented early humans. The golden rule, do to others that which you would like done to you, is a guiding principle of many very early cultures and religions that relies on being able to create a mental model for how others perceive an individual's actions. The emerging picture is one in which animal and human consciousness vary not in kind but in degree. Despite these new discoveries, the mental distinctions between humans, higher primates, and other animals remain murky and may always be so.

THE EVOLUTION OF HOMINIDS

Hominid fossils dating as far back as 4–5 million years correlate with a slow, overall increase in mental ability. Although the advantages slowly accrued, the picture of early human evolution is one of numerous developments, many of which led to dead-ends. In this sense, there seems nothing special about human evolution when compared to other mammals or animals.

Upright bipedal walking is now regarded as the crucial behavioral and anatomical change that brought early African tree-dwelling hominids down to the savannah. Fossilized footprints left in volcanic ash around 3.6 million years ago million years ago show differences in bone structure consistent with an upright human form. Along with climate changes that favored grassy woodlands, upright early man gained several advantages in the developing environment, ranging from better visibility of predators and prey to easier hunting of large-bodied grazing animals and better control of body heat despite the high volume of blood required to power a large brain. An upright posture frees up the arms for carrying weapons, food, or belongings, and for more easily reaching higher fruit from bushes.

Widespread support exists for the emergence of modern hominids from Africa. The "Out of Africa" hypothesis is strongly supported by genomics, fossil evidence, and current knowledge of geographic changes. Unlike the Neanderthals in Europe or descendants of *Homo erectus* in northern China, the human inhabitants of Africa suffered from few temperature and sea level changes.

Australopithecines are among the earliest hominins who were able to walk on two feet and had a cranium of similar size and shape as modern apes. Although skull size is not a measure of cognitive ability, the increased mental processing in the large frontal lobe of later hominids, correlates with a projecting forehead which is not present in the *Australopithecines*. Various sub-species of *Australopithecus* spread throughout Africa, and then became extinct about 2 million years ago. From roughly 2.3–1.4 million years ago the fossil record shows the presence of *Homo habilis* with a larger brain capacity and a correspondingly greater level of manual dexterity. *Homo habilis*, or "handy man" made the first recognizable tools, stone flakes used for cutting. Interpretation of the flake patterns support the hominids being right handed, a trait which correlates with asymmetry between the two sides of the brain. Archeologists reproducing these tools show that a surprising amount of skill is needed to strike one stone against another to produce usable flakes.

Perhaps most significant is that the tool maker must have mentally conceived an image of the final form before crafting the flakes out of formless rock, a level of cognitive sophistication seldom observed in the animal kingdom. Some of the best rocks for making these tools appear to have been carried to other sites, presumably in anticipation of using them later. Across the span of roughly a million years on three continents, early hominids were making the same stone axe leading ultimately to the teardrop hand axe, the "Swiss Army Knife of the Paleolithic" which persisted until around 10,000 BC.

Homo habilis was surpassed about 1.8–1.3 million years ago by *Homo ergaster*. The latter exhibited a greater tool-making ability and the first evidence for a modern family structure where fathers provided food, and protection for mothers and children. *H. ergaster* was more like modern humans in body structure and likely lived in hunter-gatherer societies similar to those still extant today.

Homo erectus, 1.8–0.3 million years ago, had a small increase in brain size relative to *H. ergaster*, and a greater degree of tool sophistication. *H. erectus* mounted stones on wooden handles for use as axes and appears to have tamed fire around 800,000 ya, although the precise date is the subject of some speculation. Fire allowed cooking which, although diminishing nutrients and vitamins, reduces the time and energy needed for digestion and leaves more time for social interaction.

Perhaps the most famous of the more recent hominids is *Homo neanderthalis* that lived from 150,000 to about 30,000 ya. Neanderthals were relatively successful in living through several ice ages despite fossil evidence routinely indicating poor diet and fractures from difficult lives that saw most die before the age of forty. The cognitive ability of the Neanderthals is debated; the distinctive skulls are sloped back with less forehead for the frontal cortex where much mental processing is performed. Neanderthals made tools and buried their dead but without the elaborate grave goods associated with modern humans that abruptly started about 40,000 years ago. Trying to parse whether Neanderthals thought in the same way as modern humans is a difficult question that is likely to be debated for some time because only physical clues remain in witness of immaterial mental cognition.

Neanderthals, having adapted over 100,000 years, were abruptly displaced by modern humans within about 13,000 years. Whether Neanderthals were physically unable to compete for resources or were wiped out by the emergence of more intelligent *Homo sapiens* is debated. Modern genomic analysis finally established an intermingling between modern humans and Neanderthals in Europe and Asia, but not Africa. One estimate of the Neanderthal gene input into the *Homo sapien* gene pool is 2 percent, though the details and influence of the gene sharing are not yet fully understood.

Human fossil remains occupy branches on an evolutionary tree rather than a progression from the earliest hominid to modern *Homo sapiens*. Neanderthals occupy one branch that diverges from modern *Homo sapiens* and stops. Hominid development written in the fossil record is one of evolutionary exploration in which multiple new hominid species emerge only to subsequently perish. Human evolution has been anything but a slow steady progression despite only one hominid being present in the world today. Only by viewing all hominids can the evolutionary tree be seen to point to a last common ancestor whose species is no longer alive today. The line of hominids producing *Homo sapiens* has disappeared but the influence of this one species on the world has not.

THE EMERGENCE OF MODERN MAN

Forty thousand years ago the first anatomically modern humans were settling in Europe. Like Neanderthals, these *Homo sapiens* were big-game hunters and nomadic or semi-nomadic. Despite the hostile ice age environment they developed a sophisticated, artistic culture captured in the decoration and design of tools and the fantastic cave paintings found at Altamira, Spain, and Lascaux, France. The cave art is speculated to have been made by shamans whose visions and potential religious symbolism were inscribed on the walls and ceilings of caves far underground. Unlike Neanderthals, these first modern humans left evidence of making and using tools, the weaving of cloth and use of skins for clothing, making jewelry, good communication, burial rituals, and even a calendar.

What happened 40,000 years ago that triggered the transition to modern humans? Some speculate a "phase transition"—a small genetic change that had a much greater impact than one would expect and which allowed the barrier to modern cognition to be transcended. Similar transitions are thought to have occurred during the Cambrian explosion, such as when animals started walking on land. Theories vary, although all agree on a fairly dramatic mutation.

The cognitive change in these modern, early hominids was followed by a fairly rapid series of cultural changes. Around 34,000 years ago people began making bone points for spears, and by 18,000 years ago javelins appeared, greatly increasing the speed and accuracy of the lethal projectiles. Bows and arrows emerged slightly later. From about 20,000 years ago, some archeological sites show an increase in the numbers of people living together and the presence of tools made from materials quite remote from where they were found. Modern genomics shows that these humans were not the end of an evolutionary progression. Biological evolution accelerated once farming began around 10,000 years ago and continues today.

Farming had both cultural and genetic influences. Domesticating cattle resulted in northern Europeans producing new enzymes for digesting milk. Accompanying this genetic change was a cultural change from hunter-gatherer to domesticating animals to convert grass to produce protein which freed time for individuals to develop the skills characteristic of modern civilization.

Bartering emerged during the development of early modern humans, a distinctly human activity engaged in by all human tribes. Archeological evidence suggests that trade has been occurring for thousands of years with precious stones, metals, and shells traveling hundreds of miles to other people groups. While animals engage in reciprocity, such as delousing, none barter. Efforts to teach chimps to barter have failed.

Trading requires some individuals to become specialists and others who are willing to trade. Trade provides specialized items not otherwise easily obtained and saves time for other pursuits. Trade and innovation appear to have gone hand in hand. The development of trading networks may indicate the ability of strangers to negotiate through different people groups, perhaps facilitated by a recognition of the mutual benefits of cooperation. Women may have played a key role in developing the trust required between trading groups. Marital relationships in which women move away from their relatives to form new family units with men from a different region would establish new family ties and the trust necessary for trading.

LANGUAGE

Language has become a litmus test for differentiating people from animals. Chimpanzees have a different vocal apparatus from humans and different physical connections in the brain area normally dedicated to language. Recent gene studies have focused on the gene known as FOXP2 which is tightly linked to language ability. Chimpanzees have a gene that differs from human FOXP2 by two amino acids, which indicates a potential site of variation. FOXP2, while not found in chimpanzees, has been found in Neanderthals, implying the potential for language. Whether Neanderthals communicated in the same way as people do currently is a very different and strongly debated question.

Humans speak from their brains in the sense of having sophisticated cognition that is relayed through the vocal chords. Why humans developed a long larynx subject to choking remains something of a mystery. Evidence for the development of language is difficult to find and is often inferred from skeletal features. Typical mammals have a short larynx, which prevents choking but also speech. Changes in the base of the skull are seen in adult modern humans whereas apes have a relatively primitive vocal tract and flat skull bases. Although not a perfect indicator of speech capability, the flatness of the skull base provides some guidance. The first hominids with skulls that could accommodate a modern kind of vocal track, *Homo heidelbergensis*, did not leave the rich legacy of the later *Homo sapiens*.

Extensive effort has been expended to train higher primates to communicate as a model for understanding the development of language. Efforts to teach language to chimps in captivity have achieved several milestones. Chimps can learn to interact and respond to basic queries, such as "color of apple." They can learn word strings and put them together. The words they learn are those with a sensory component whereas abstract concepts like time and metaphors appear beyond them. They do not ask questions. Extensive research indicates that apes are not capable of creating sentences despite understanding symbols and being able to imitate their trainers. Their communication can be very expressive but they do not link symbolic sounds together in ways close to language.

In the wild, chimps use a series of vocalizations to relay warnings. Playing back prerecorded chimpanzee screeches for "leopard warnings" causes chimps in the field to scatter to the trees. "Eagle warnings" cause them to scan the skies and the "snake" call makes the chimps gaze on the ground. Chimp expert Jane Goodall believes that communication among chimps involves sounds intimately associated with emotional states. In this sense their ability is like that of parrots, dolphins, finches, and killer whales. They combine calls but do not develop the ideas into a series of thoughts. Virtually absent are the rules of grammar and syntax central to human communication leading most researchers to agree that a huge gulf exists between human language and communication among primates.

In *The Ape and the Child* the Kelloggs describe their experiment in which they raised a seven-month old chimp along with their ten-month old son. During the first few months the chimp was "considerably superior to the child in responding to human words."[1] However, by two years of age humans have usually amassed a command of language that continues to grow at an accelerating pace different in kind from that of a chimpanzee. *The Ape in our House* and *Lucy: Growing Up Human* describe similar experiments with chimps who learn some behaviors and communication skills, but dramatically fewer than those of human peers.

---

1. Kellog and Kellog, *The Ape and the Child*.

The interpersonal skills of young humans and apes are similar up until the age of four, at which point ape development essentially ceases while humans continue.

While apes have complex social behaviors and communication modes, human language appears to represent a quantum jump in communication from any other living entity. Learning language during childhood correlates with maturation in the brain and diminishes strongly after puberty. Children rapidly acquire the ability to assemble longer and more complex sentences whereas the ability in chimps tails off after putting a few symbols together.

Language provides an efficient mechanism for processing the autobiographical memory that forms the basis of personal identity. Developed language facilitates complex mental representations such as the pursuit of abstract goals like justice and the ability to predict consequences. Children develop marks of human rationality as they consider their actions and, after examination, decide whether they should act differently.

The evolutionary transition from pre-hominid to "wise man," *Homo sapiens*, occurred over the same time frame as language and complex self-reflection developed. Sophisticated animals possess a capacity for an awareness of others but not to the same level as in humans. Humans have a particularly high capacity for understanding how actions will be interpreted by others. Higher order consciousness in humans is significantly aided by symbolism which allows complex mental constructions to be evaluated, envisaged, and acted upon or not. The ability to develop models of what others might be thinking provides one measure of cognitive development with "orders of intentionality" being roughly correlated with hominid evolution. Humans are capable of understanding five levels of intentionality: I know (1) that you know (2) that I know (3) that you understand (4) that God can see your thoughts (5).

Language provides a means by which individuals communicate abstract concepts from one mind to another. The enhanced language capacity of humans is the basis of a deep, rich, dimension of relatedness with people, and potentially God, that is unavailable to nonhuman primates. People are able to understand the external world, formulate thoughts, communicate them, and take action on them. No other system, biological or man-made, no matter how intelligent, is able to replicate this process.

HUMANS

Humans are the most culturally sophisticated species in earth's history. People distinguish themselves from animals by the extent to which they modify the world around them. Language, dexterity, visual skills, sociability, self-reflection, and intelligence set the human *person* apart from any animal. Gene sequence mutations allow scientists to compare the relationship between *Homo sapiens* and every other organism for which data is available. Genetic comparisons between early hominids dovetail extremely well with the evolutionary tree derived from fossils.

Tracking chromosomal DNA provides a means of following genetic changes as new species evolve over time. Longer time frames can be probed by examining the DNA responsible for encoding the mitochondria, the cellular power houses found in plants and animals. Mitochondrial DNA is directly inherited from the mother and is relatively unchanged over vast periods. Mitochondrial DNA is virtually the same for the entire human population. Tracing the occasional mutations back in time leads not to one mitochondrial Eve, but to a small population of early humans.

Human DNA is very similar to DNA obtained from Neanderthals and progressively less similar to chimpanzees, apes, mammals, and other animals. Humans share 99 percent of their genome with chimpanzees and 98 percent of their DNA with gorillas, and yet the 1–2 percent difference is significant. After all, humans are determining the fate of chimpanzees and not the reverse. Although 1–2 percent doesn't sound like much, this represents around 30 million differences in the amino acid sequence. Just as important as gene similarities are non-coding regions of DNA that regulate developmental genes. In the regulation of genes and their copying, humans and chimpanzees differ by 6 percent.

Some of the gene differences are more important than others but should be put in the perspective that people share 99.9 percent of their DNA with each other. The 0.1 percent that distinguishes each unique individual is important; basketball superstar Michael Jordan and the musical genius Mozart share 99.9 percent of the same DNA and yet their talents are completely different.

Behavior is not determined by genes alone but involves an element of personal choice. The choice of whether or not to smoke provides some insight into the way genetic factors can influence behavior. Teenage smoking is intrinsically linked to sharing an emotional experience with friends through a counter-cultural ritual that expresses identity. Smoking is not genetically encoded but is more prevalent among people who are gregarious, crave excitement, are impulsive, take risks, are aggressive, and are sexually active. Among numerous theories accounting for the correlation between smoking and depression, there is some evidence that nicotine mimics the action of key neurotransmitters in the brain. Smokers may be knowingly or unknowingly improving their mood through self-medication.

Somewhere in the evolutionary process there occurred a remarkable transition in which brains, and particularly the human brain, began storing more information than is encoded in DNA. Humans create ideas that are intangible, living solely in the mind—stories, imaginary worlds, spiritual and religious beliefs, and complex mathematics. Why the human brain evolved to a level of complexity far beyond that of other primates is something of a mystery. One intriguing theory posits that social interactions caused humans to have a much greater neocortex than any other primate. There is a direct correlation between primate group size and cognitive ability. Powerful brains are required to track increased social complexity that comes with increased interactions. The larger the group, the more individual social contexts must be remembered along with knowing how those individuals interact with each other.

For humans, the maximum number of people with whom a genuine social relationship can develop is about 150. The 150 cap fits with the primate-brain correlation, the sizes of many past and present primitive people groups, and the size of many groups, such as military units, that need to work together as functional units. For example, members of the Hutterite religious group live in community but follow a rule of forming new colonies when their numbers broach the 150 limit. From experience, their leaders find that above roughly 150 people the close-knit fellowship is not maintained.

HUMAN DEVELOPMENT AND RELIGION

Homo heidelbergensis controlled fire, constructed shelter, and made tools, but arguably lacked the kind of innovation, symbolism, and religious inclinations apparent in later hominids. In some instances, the more recent Neanderthals exhibit burial practices with greater sophistication than earlier hominids, but the extent to which this indicates a strong religious belief is unclear. Modern-day hunter-gatherers view themselves more as a part of the ecosystem with attendant religious beliefs whereas agrarian cultures tend to view themselves in opposition to nature, storing food and planning for each season.

Spiritual inclinations require envisaging a life that transcends immediate experience. Although the interpretation of grave goods and burial practices is a source of contention, there is widespread agreement that some early modern humans, *Homo sapiens,* gave far more complex treatments to their dead than any previous hominids. Archeological findings dating from around 40,000 years ago show *Homo sapiens* making tools, paintings, small sculptures, and elaborate grave goods. The presence of decorative shells among grave goods, in some cases perforated and sown or laced together, could be a cue that these early modern humans considered death more than just the cessation of existence. At a deeper level, grave goods suggest an empathy with the dead and possibly a spiritual acknowledgement of a transition from a physical life to a spiritual life.

The rise of religion over time has elicited radically different interpretations. At one extreme religion is seen as a perpetuating virus imparting survival value by describing dangers to be avoided and character traits that might aid small populations to get along well. Indeed, recent medical research shows that religious beliefs alleviate depression and anxiety and give individuals longer, happier lives. Historically, the religious ideas of early groups of humans were thought to provide explanations for powerful and unusual natural events such as an eclipse, a lightning strike, or an earthquake, by linking them with spiritual events.

Recently, children have been suggested to be predisposed to religious belief. The inclination toward belief might develop naturally as a consequence of the way children learn about the world. In the mind of a child, God begins as an unseen thinking being more or less like a person, but over time becomes transformed into an all-knowing, all-powerful being. Experiments with three-year-old children show an inability to differentiate between the mortality of a person and of God. Young children intuitively take human traits and project them onto God, refining their ideas as they age. A child's perception of God is easily developed because the assumptions are readily accommodated in the natural projections of the mind. Interestingly, children younger than four have difficulty understanding that beliefs can actually be false or differ from person to person. Just try telling a four-year-old that Santa is not real!

By five, most children understand that people may believe something that is not true, but they also think that God is not fooled even in cases where people are. For example, kids over five presented with a cracker box containing rocks initially believe that the box contains crackers. After being shown the real contents they then believe that if their mother saw the box then she would think the box contained crackers but God would know the box contained rocks. Studies by developmental psychologist Jean Piaget, suggest that children younger than seven assume that parents and God can do anything. Trauma occurs when children subsequently learn that parents cannot change everything, leaving God as the only being who can.

Identifying humans as different in kind from primates and/or early hominids aligns with the historical view of God endowing an individual line of humans with spiritual knowledge. Scientific discoveries continue to challenge the view that humans are materially different in kind from primates, a fact compounded by debilitating diseases capable of reducing mental cognition to levels similar to that of advanced primates. Greater understanding of cognition has caused Christians to reevaluate the early chapters of Genesis with a renewed focus on what sets people apart from animals. The answer lies in the relationship that God establishes in Genesis: God walked with Adam and Eve in the garden and knew them by name.

NATURAL EVIL AND GOOD

People seem to have an inherent understanding of good and evil and, as in many epic movies, believe that the dark side will ultimately be defeated. Good is experienced as that which is welcomed, enjoyed, and sought after, whereas evil is that which is disliked, feared, and resisted. Theologically, good is whatever aligns with God's purpose for the world while evil is in opposition to God's purpose, usually experienced as sin or suffering.

Scientists have sought mechanisms, genetic, instinctual, or otherwise, for good, virtuous characteristics such as love, sympathy, and care of others, and also seek to identify the reason for traits that are selfish, bad, or evil. Many acts in nature are universally understood to be immoral: infanticide practiced by elephant seals who gang up on a young pup and hyena twins who fight until one dies. A minimalist objection to infanticide being immoral is that the group population would be endangered if practiced rigorously.

The field of sociobiology, emerging around the mid-twentieth century, sought explanations for social behavior that rest solely on biological evolutionary mechanisms. Why do some animals, such as monkeys, alert others to the presence of predators by providing warning calls that could endanger themselves? And why do dolphins support injured companions at the surface to prevent drowning, without any apparent benefit to themselves? Sociobiology explains social behaviors such as altruism, aggression, and nurturance as providing genetic advantages.

Reciprocal altruism may explain costly acts of helpfulness in terms of future potential benefits if the favor is returned. Rescuing a companion in trouble or sharing food or resources might later be returned in kind at a time that saves an individual's life. In this sense, altruism has a different meaning than commonly encountered; this type of altruism is not a self-renouncing action but a genetically determined behavior in which organisms behave to maximize gene replication.

Kin selection theory has been advanced to explain a tendency to help kin survive based on an increased likelihood of gene survival, even when the outcome is fatal or costly to the individual. Alarm calls by squirrels and prairie dogs alert others to a danger while calling attention to themselves. Often the frequency of the alarms is higher when kin are close by than when the companions have no close genetic ties, indicating an animal bias to protect those who are closely genetically related. Social insects such as ants illustrate cases where workers are sterile and yet care for the community. Species that rely on cooperation show group loyalty and a tendency to reciprocate help. The impulse to help, particularly among kin, confers survival value because of an expectation for reciprocity. At root, the explanations of sociobiology seek after mechanisms to describe virtuous behavior that has long been considered the purview of religion.

Computer programs have been developed that use games to determine when to help others and when to be selfish. The model advises cooperation on the first move and on every successive move to repeat the opponent's last move. This strategy has been coined altruism with teeth, a non-genetic strategy that allows cooperation to evolve. Most people experience altruism as a choice, not as a genetic imperative. From a purely biological perspective, pursuing selfish action is usually in a person's best self-interest but falls short of the altruistic actions of modern-day heroes who labor and give their lives for the betterment of others.

Altruistic behavior in primates is sometimes very similar to that of humans. Binti Jua, a female western lowland gorilla in the Brookfield Zoo, rescued a young boy who fell into the enclosure and then carried him to an access entrance where zoo personnel could retrieve him. Scientists argue about whether this was because of human training, a natural tendency to empathize, or true animal altruism.

Complex experiments are being designed to probe just how much animals do react from instinct, learning, and empathy. Rhesus monkeys refused to pull a food-delivery lever when they could see that doing so caused a companion to receive an electric shock. Analogous experiments with rats and apes reveal similar responses, although with different levels of intensity in each species. What is currently unclear is whether the animals' refusal to shock their companions is due to an aversion to seeing stress in others, personal distress caused by administering a shock, or from a compassionate desire to help their companion avoid pain.

Natural processes resulting in good and bad allow individuals to learn from experience. Higher animals learn where to find natural good; food, drink, and companionship; and also gain knowledge about what causes pain, poor health, and loss of life. Animals use this knowledge to search for food, be aware of potentially dangerous situations, and maintain a keen awareness of predators. Only if some fawns get caught in forest fires can other deer learn how to prevent their offspring from being caught in the same way. A deer caught in a forest fire suffers and dies but provides an example to other animals capable of understanding how to avoid the same situation. Although the death is a tragedy, the possibility of death allows for good to come of the situation through knowledge given to promote others' survival.

The possibility of evil allows the exercise of a greater good in response. Good, deep, right relationships with others are only possible if a real chance exists for a fractured relationship to also be a viable outcome; the possibility of good is intrinsically linked to the possibility of evil. Philosophically, the idea is termed the "higher-order-goods defense," which argues that a world with the possibility of evil has merit by allowing the free choice of good acts in the face of bad states. A person suffering physical pain has the choice of patient endurance or complaining. In turn, a friend can show sympathy or be callous. The good or bad action chosen when facing natural evil provides opportunities for further choices of good or evil. A patient sufferer can accept sympathy and share the experience, or a friend might moderate a sufferer's complaints by showing the good of developing patience. The choice to patiently endure pain, or any suffering, can forge a virtuous character in ways that profoundly surpass genetic influences.

A real choice between good and bad requires a desire for one action coupled with the personal ability to exert control. Over time, exercising choice forges character. As freedom and responsibility increase, so too the negative impact of bad choices escalates and with more choices, the likelihood of a bad choice increases.

MORALITY

Morality is the code of values that guides choices of right and wrong. Scientists use observations to understand moral choices, whereas religions see moral rules as coming from God. The two approaches are not necessarily mutually exclusive because a world created by divine inspiration from a good God is expected to reflect an orientation toward good and a repulsion toward evil. A universe governed by simple laws of nature will allow individuals to discover simple forms of good and bad. Contact with hot objects causes pain whereas eating chocolate is pleasurable. Eating food is good for the body but eating too much is bad for the body—and can deprive others. Provided that experiences are closely connected with their consequences, an individual can learn from, and experience, the consequences of their actions and divide their actions on this basis into good and evil.

Developments in cognitive science are providing insight into how moral decisions are made. On one level, the human brain makes some decisions almost automatically in response to an enormously diverse set of conditions. On a much deeper level, other decisions are made with considerable thought, such as solving difficult mathematical questions. Having an automatic response unit provides a powerful system capable of responding to diverse input, sometimes in ways that individuals themselves are unaware of. For example, posting a picture of a pair of eyes above an office coffee bar operating with an "honesty box" afforded, on average, contributions three times greater than on weeks with a flower poster. Evidently the symbolic reminder of being watched encourages virtuous behavior.

Advances in neuroscience have helped to uncover the mental mechanisms of reflective decisions. Mirror neurons are activated in primates when they see, and then share, the same emotion seen in others. Seeing an emotion or action in one individual causes the same sensation in the observer. Distress at seeing pain in another person is a reflex. The effect seems to cross species; a monkey being monitored by MRI for brain activity, saw a scientist enter the testing area, take a nut, and begin to remove and eat the contents. Scientists found that the monkey's MRI pattern was indistinguishable from that obtained when the monkey performed the same action.

These experiments suggest a biological component to empathy, but they do not demonstrate that morality stems solely from the innate workings of the brain. Distilling all human actions down to deeply selfish origins pertaining to the survival of the individual or their genes disregards people's intellectual ability and glosses over the mechanisms of mental processing in the human brain.

Morality among early people groups may have been fostered through a concern for conflict resolution, cooperation, sharing, and ways to regulate individual conduct for the overall good of a group. Humans have long banded together for safety and for warfare, suggesting that if morality has an inherited component, this may have developed from base desires that were far from moral. Selfish behaviors like stealing, cheating, and promiscuity likely diminish group cooperation and disrupt benefits such as group hunting, bartering, and tending of the sick. Greater cooperation among people may have clarified the distinction between right and wrong and fostered moral behavior. In this sense, an understanding of good and evil might emerge naturally from mutually beneficial group dynamics.

An evolutionary ancestry provides an explanation for the malevolent animal-inherited tendencies that plague people's lives. Lust, aggression, and dominance are behaviors easily linked to innate sexual and animal instincts. Although some inherited animal behaviors are not bad in themselves, they require proper direction and control for good purposes. A tension exists between natural desires inherited from an animal ancestry and individual choice. Religion explains this tension as a spiritual problem known as original sin, perhaps the only empirically verifiable religious doctrine.

ADAM AND EVE

The Genesis story describing the creation of Adam and Eve is very different from the story of modern man told by science. Whereas the first three chapters focus on the creation of one couple, geneticists identify a "mitochondrial Eve" as one woman among about 10,000 early modern humans. Through this one small group, a dominant gene line spread through all people alive today. Simple population genetics based on a literal Adam and Eve living about 10,000 years ago at the beginning of the Bronze Age described in Genesis, would be unable to populate the world to the extent of 7 billion people alive today.

Several interpretations have been proposed to accommodate biological evolution with an interpretation of Genesis. While no one interpretation has been universally accepted, the different approaches have helped address apparent inconsistencies between scientific insight and the stories of Genesis.

The first eleven chapters of Genesis are written in a distinctive style. Many modern theologians regard Genesis as a true mythical description of man's fall that uses phenomenological rather than literal language. The formation of the first people, the presence of good and evil, and the spread of language are told through a series of stylized stories that focus on humanity's condition and a broken relationship with God. Even the first mention of Adam ("human") and Eve ("living one") in the Bible uses symbolic rather than individual titles. The name Adam is a pun on the Hebrew word for earth, equivalent to the English "earthling" from the "earth" or "human" from the "humus"! Only later, in the genealogy of chapter 5, is Adam mentioned by name as an individual.[2] Interpreting Adam and Eve as symbolic individuals allows for earth to have other inhabitants, as hinted at in other parts of Genesis (e.g., Cain and Abel's wives, the people that Cain was worried might kill him, and the inhabitants of the city Cain was building[3]).

One approach identifies Adam and Eve as literary types, symbolic of the ancestral colonies from which modern man arose. A challenge with this typology is to understand later biblical passages that seem to compare a literal Adam with Christ. An alternative to the literary typology, is to consider Adam and Eve as representative of a group of early hominids who became modern humans under God's divine plan.

2. Gen 5:1–5
3. Gen 4:17

Homo divinus describes the first pair who were spiritually alive and in relationship with God. In this view, Adam and Eve may have been real people living in Neolithic times, who spread the new revelation of God among an existing community. The point of the story is the spiritual understanding of God and the ensuing relationship rather than genetic propagation and population growth. None of the approaches to reconcile the story of Adam and Eve with the newly acquired genetic information completely accommodates the early chapters of Genesis as an inspired, narrative theology for the early Hebrews.

Just as modern science has stimulated a renewed interpretation of Genesis focusing on the relationship between Adam and Eve and God, so too have theologians begun to interpret the Garden of Eden as something other than paradise. Historically, the Garden of Eden was seen as a paradise with no prior suffering, no volcanoes or earthquakes, no carnivorous animals, no pain, and no death. Such an interpretation would require massive biological changes at odds with what is known about the world through science. A symbolic reading of the Garden of Eden as a fertile, well watered place of abundance fits with God giving Adam and Eve an ideal, and realistic, place to live. Such an interpretation fits clues in the later part of Genesis that suggest that Adam was mortal, just like people today, unless he "reach out his hand and take also from the tree of life and eat, and live forever."[4]

MORAL EVIL

Genesis 3 describes the introduction of evil. A snake tempts Eve to eat from the tree of the knowledge of good and evil whose fruit God specifically prohibited.

> Now the serpent was more crafty than any of the wild animals the Lord God had made. He said to the woman, "Did God really say, 'You must not eat from any tree in the garden?'" . . . When the woman saw that the fruit of the tree was good for food and pleasing to the eye, and also desirable for gaining wisdom, she took some and ate it. She also gave some to her husband, who was with her, and he ate it. Then the eyes of both of them were opened, and they realized they were naked; so they sewed fig leaves together and made coverings for themselves.[5]

4. Gen 3:22
5. Gen 3:1, 6

The pair's disobedience caused self-realization and a dramatic relational change; shame, fear, blame, and alienation result. Adam and Eve were naked before eating the apple, but now they realize their naked state. The relationship between Adam and Eve, and between them and God, was corrupted by disobedience.

The tree of the knowledge of good and evil symbolizes the choice of people to obey or disobey God. Evolutionary biology, which links primitive hominids with modern humans, provides a complementary explanation of personal choice based on biological ancestry and cultural heritage. People's inherited biological instincts cause an inclination toward preservation at others' expense, which religion interprets as a constant falling away from the divine ideal of caring for others. The third chapter of Genesis explains Adam's sin-severing relationship as more than physical death; life with God has been corrupted. The remaining books of the Bible describe the relationship of God with the chosen people of Israel and the ultimate restoration of relationship with God through Jesus.

The Garden of Eden story raises the difficult question of how evil arose in a world created by a thoroughly good God. Augustine (354–430 AD) developed the most influential theology accommodating God's goodness and omnipotence with the existence of evil. He framed evil as the absence of God, a privation that can arise in nature through disease, accident, and death. Evil is a direct affront to God's love expressed as an inexhaustible, and ongoing creation that begins by filling the world with all possible creatures from the greatest to the least. Theologically, the privation view of evil avoids any divine responsibility because evil only exists as the malfunctioning or disorder within an essentially good creation.

Irenaeus' (130–202 AD) theodicy, the defense of God's goodness despite the presence of evil, focuses on the relationship between God and his people. Irenaeus distinguishes between the *image* of God, the nature of intelligent beings capable of relationship with God, and the *likeness* of God, which is the perfecting of an individual's nature through spiritual development. People are endowed with moral freedom and bear the image of God, but the relationship requires a spiritual development to draw each person toward the perfection intended by God.

Understanding the first chapters of Genesis as a story of first beginnings anchors a part of evil in spiritual immaturity. Adam and Eve can be viewed as naïve innocents who succumbed to the serpent's seduction. People's awareness of God lags behind their developing physical nature, resulting in sin, guilt, and the need for redemption. Sin arises both in each individual and through societal influence because society is corrupted by the presence of sinful individuals. God redeems sin and evil, using the situation to serve the ultimate good purposes of God. Sin and redemption are like an accident repaired by surgery in that the surgery repairs damage and saves lives. Sin is utterly opposed by God, but is allowed to occur because God's grace is able to freely draw an individual toward redemption.

People are made as imperfect beings that are able to be brought to perfection through divine grace. Each person is created in the image of God and through grace can attain the likeness of God as revealed in Christ. Spiritual maturity comes through confrontations with evil arising from conflict, pain, and suffering that are redeemed through God's grace. In this sense, the image of God imprinted on each person is translated into the likeness of God through a struggle with evil. People who meet and master temptation transform their souls in ways that come from the investment of personal, and often costly, effort. People move toward perfection through a haphazard adventure in which individual choices ultimately allow progressive fulfillment, or not, of God's purpose. The process is not an inevitable evolutionary process, but depends on individual choice.

At times evil is defeated by good: sin ends in repentance, danger evokes courage, difficulty produces patience, and temptation produces moral steadfastness. At other times wickedness multiplies, suffering weakens character, and hope changes to despair. But Christians believe that the ultimate victory of good is ensured because of the ultimate triumph that has been won through Christ's atonement on the cross. The pain of life, sin, and suffering are not removed, but Jesus' atonement provides a victory for all that death cannot touch.

Harmonizing biological evolution with biblical theology requires the merging of the long, slow development of man with a spiritualization resulting in individuals capable of a personal relationship with God. A minimal level of cognition and an environment free of coercion is required for the development of free and self-directing individuals capable of choosing to have a conscious relationship with God. Personal choice and autonomy in a good world created by God creates a paradox. Only if God places individuals at a distance from himself can people have a voluntary choice. The human environment needs to have God sufficiently obscured that life can be lived without God to allow for real choice, while having sufficient elements of God's reality exposed to allow individuals to be drawn toward God through signs of divine presence. A hallmark of this world is ambiguity. Sufficient ambiguity must exist for some to discern the hand of God moving within history while allowing others to rationally live without acknowledging a divine presence.

Evolutionary theory proposes the arrival of the first people in a close organic relationship to the world. Modern humans emerge into a world with a choice. They can view reality as pointing to a divine guiding presence in and through creation or not. The biblical account has God in the garden among the first potential believers, who were summoned from fertile evolutionary dust, and are able to freely choose to maintain a relationship with God or not. Man arrives on both scenes at a distance from God.

Ultimately, the existence of evil inescapably rests on God. There is no-one else to bear the responsibility if God brought the universe into existence out of nothing. Despite God willing the universe into being and knowing the course of creation, each individual is responsible for their own sins. God knows sin, suffering, pain, and evil will occur and lovingly offers a pathway to restoration for each person. The foundation of God's rescue is Christ's death and resurrection: O fortunate crime that merited such and so great a redeemer (quoted from the Easter Paschal Vigil Mass Exsultet). God allowed creation to fall because redemption was planned from the beginning.

CONCLUSION

Evolutionary and religious beliefs are often in tension. Despite the wealth of scientific support for evolution and the vast number of people with religious experiences, few try to develop a coherent worldview encompassing biological science and religious writings. Consequently, while the US produces much of the leading evolutionary science, a large portion of the American public believe that God made and populated the world in a few thousand years.

Advances in biology, particularly through genomic comparisons, provide a compelling argument for a long evolutionary development from primates to modern humans. The sequence is one of both proliferation and extinction before arriving at the current hominid form that might loosely be translated as thinking man. During this progression, key human qualities such as language and cognition become more prominent, as interpreted from the remains left behind by these distant cultures.

The most intriguing development of human evolution is the transition from early hominid to modern people. About 40,000 years ago a dramatic change occurred in the cognition, language ability, and religious understanding of early modern humans. While evolution explains much of *Homo sapiens'* development, characteristics like empathy, love, hope, and trust, call into question whether this is the only mechanism by which the development of modern humans has been guided. Is a strictly evolutionary process enough to explain people's appreciation for precisely those attributes that separate animals from humans? Sociobiology provides compelling explanations for human attributes, such as altruism, that directly impinge on an understanding of good and evil. However, the ability to craft beautiful art, play evocative music, and write emotional stories seems to require excessive energy without any adaptive advantage and would be expected to disappear through evolutionary pruning. Are other forces at work to draw out and enhance distinctively good human potentialities?

Perhaps the most difficult intersection of science and religion lies in understanding the triad of good, evil, and autonomous choice. Good and evil seem intrinsically linked in ways that some interpret as pointing to a divine presence, but which others discount because of the pain, death, and suffering inherent in the world. The Garden of Eden captures the many elements of good and evil in a simple story whose essence is that man's relationship with God is broken. Understanding Adam and Eve's origin as having a long biological heritage focuses attention on animal inclinations that directly impinge on relational problems. From a theological perspective, God's incarnation in the person of Jesus is the centerpiece of redemption which allows the broken relationships described in Genesis to be restored. Entering into religious relationship entails consequences because a recognition of a divine imprint requires a recognition of sin, a desire to change, and a willingness of the individual to restore the broken divine-human relationship.

DISCUSSION QUESTIONS

1. Chimpanzees have been used extensively in medicinal studies to model the influence on humans without subjecting people to the same tests. Knowing the level of sentience exhibited by chimpanzees, what should their status be for research? Should chimpanzees be given the same legal rights as humans, as has happened in Spain?

2. During the long course of human history, God provides only humans with the opportunity of salvation through Jesus, well after all other hominids are extinct. Why did God wait so long before offering the possibility of salvation? How is God's exclusivity for *Homo sapiens* justifiable?

3. Humans have exceptional mental and physical abilities. Does this capture all that being a person means or is a spiritual component an essential element of being human?

4. Modern medicine is amazingly effective in replacing organs and keeping people alive. Heart transplants are now relatively routine with more successful operations each year. Medical technology is currently exploring transplants of a human head to provide new mobility for paraplegics. After a transplant using a donor's cranium, who is the transplanted individual?

5. Genesis declares that "God created man in his own image, male and female he created them."[6] What exactly does this mean given the likelihood of the evolution of modern people from earlier primates?

6. Many children with autism do not understand why a person would want to deceive another and why anyone would believe something that isn't true. In this sense, these children would fail the test for having multiple levels of intentionality. Does this demonstrate that the theory of mind is inadequate to differentiate humans from primates? Should the definition be changed, and if so how?

7. How does a person's genetic predisposition toward an addiction influence a person's moral responsibility to conform to social or legal norms?

8. Justice is a moral concern focused on putting right some wrong, often through punishment. Something stolen must be replaced. On

---

6. Gen 1:27

what basis might justice be tempered and mercy extended? For example, how should a person be treated for stealing food for a starving family?

9. Evolution directly impinges on what being human means. Are humans animals or something more?

10. Most people prefer to think of themselves as conscious, thinking beings with the capacity to understand the world, bring pleasure to others, potentially achieve great things, and to live life with a purpose. Is there room for humans to have evolved and yet have the special status asserted by the Bible?

Further reading for "Primates, Hominids and Humans: What Makes People Human?"

1. Matt Ridley, *The Rational Optimist: How Prosperity Evolves.* New York: HarperCollins, 2010. Ridley believes that trade is an essential characteristic of modern people that fosters innovation. He has famously coined the term "when ideas have sex" to capture the evolution of new ideas that seems to occur through exchange for goods and services. Although several of his proposals build on the fringes of current theory, his overall argument for the distinctiveness of modern man is compelling.

2. Ian Tattersall, *Becoming Human: Evolution and Human Uniqueness.* San Diego: Harcourt Brace, 1998. Ian Tattersall makes the study of human fossils fascinating, putting flesh on the bones of prehistoric man. Following the evolution from apes and chimps, Tattersall portrays arguments for and against assigning character traits to early hominids. He distinctly conveys the imprecise art of understanding how much fossil evidence can tell about lifestyles many years after the fact.

3. Malcolm Gladwell, *The Tipping Point: How Little Things Can Make a Big Difference.* New York: Little, Brown, and Company, 2000. Gladwell's best-seller follows societal trends and searches for the reasons why groups of people make the choices they do. Using a variety of studies, he identifies rules that seem to universally influence human behavior.

4. Daniel, Kahneman, *Thinking Fast and Slow*. Reprint. New York: Farrar, Straus and Giroux, 2011. Psychologist and Nobel laureate Dan Kahneman unravels how decisions are made in the mind using reference to two main systems: an immediate response system that operates on autopilot, and a deeper, more thoughtful processing system that requires significant, directed thought. The focus of the book is decision making, particularly understanding why people make seemingly irrational choices.

5. Jon Cohen, *Almost Chimpanzee. Searching for What Makes us Human in Rainforests, Labs, Sanctuaries, and Zoos*. New York: Times, 2010. Journalist Jon Cohen treks through the African savanna, interviews scientists, and visits sanctuaries to interview researchers and learn what distinguishes chimps from humans. The beauty of the book lies in the attempt to gain perspectives on mental and physical attributes from experts that convey the majesty of endangered apes and clarity about their difference to humans.

6. Paul Davies, *The 5th Miracle: The Search for the Origin and Meaning of Life*. New York: Touchstone Foundation, 1999. Science popularizer Paul Davies engages the most perplexing questions on the origin of life in an easy-to-read style. Davies weaves possible biological theories together with life on Mars, Panspermia, and other theories that build on his earlier writings on cosmology. Davies is one of the finest, fairest writers, with a poetic style that keeps the mystery of life and engages a few of life's grand questions along the way.

7. Denis Alexander, *Creation or Evolution. Do We Have to Choose?* New York: Monarch, 2008. Denis Alexander covers evolution from creation to intelligent design. The focus is on biological evolution and a parallel exposition of biblical passages. Chapters focus on several different issues, including the identity of Adam and Eve, the biblical understanding of death, the fall, and natural evil.

8. Chris Stringer, *The Origin of Our Species*. London: Penguin, 2011. Chris Stringer, one of Britain's foremost experts on human evolution evaluates modern theories of hominid evolution. Particularly interesting are theories of mind, the development of language, and speculations on what caused *Homo sapiens* to differ from other hominids.

9. Frans B. W. de Waal, *Good Natured: The Origins of Right and Wrong in Humans and Other Animals.* Harvard: Harvard University Press, 1996. Frans de Waal uses his years of primate research to sketch out natural tendencies capable of eliciting moral behavior. While not a sociobiologist, de Waal does believe that much moral behavior can be explained by innate mechanisms rather than through learned behavior. The book is particularly valuable in drawing out differences between primates and man.

10. Ian Tattersall, *Paleontology: A Brief History of Life.* Conshohocken, PA: Templeton Foundation, 2010. Tattersall races through the evolution of life from the earliest rocks and fossils to the arrival of man a few billion years later. In the final chapters of the book, Tattersall, a curator at the Museum of Natural History in New York, follows the development of modern hominids by focusing on what makes *Homo sapiens* unique.

11. Alvin Plantinga, *God, Freedom, and Evil.* Grand Rapids: Eerdmans, 1989. A classic study of how a truly good God can be reconciled with the presence of evil. Plantinga reviews and evaluates the main theological arguments in a reasonably accessible approach.

12. Ian Barbour, *Nature, Human Nature, and God.* Minneapolis: Augsburg Fortress, 2002. Barbour addresses some of the more difficult recent topics at the interface of science and religion. Particularly insightful is the discussion of evolution and genetics in which Barbour argues that human choice, not genetics, governs human behavior.

5. Jesus Christ: Prayer, Miracles, the Causal Joint, and the Resurrection

HISTORY IS DIVIDED INTO two parts by the birth of the world's single most influential person: Jesus Christ. Jesus wrote no great work, never commanded a nation, and created no political movement, and yet, two millennia later, he is claimed as God by a third of the world's population. Who exactly is Jesus Christ, what was his message, and is he really God in human form?

Jesus of Nazareth lived two thousand years ago in a relatively unimportant part of the Roman Empire that was a constant source of friction for Rome. Details of his childhood are scant, with most writings focusing on a three year period of Jesus' life from thirty to thirty-three. Despite a lack of detail, virtually all historians of antiquity believe that Jesus was a real person. Apart from the four Gospel accounts in the Bible, historical accounts of Jesus are limited to a few documents that describe where Jesus lived and his execution as a criminal under the authority of Pontius Pilate, the Roman Prefect of Judea.

The four Gospels provide the fullest description of who Jesus is and the divine message he brought for the world. The Jews believed that, after an absence of 400 years, God was again moving among his people and that the promised Messiah would appear. Mary, Jesus' mother became pregnant through the power of the Holy Spirit. How she became pregnant, and where the extra set of chromosomes came from is not clear. What is clear is that Jesus was different from anyone who lived before or since. Indeed, Christians believe that Jesus is both human and divine.

Jesus' birth introduces two prominent themes woven throughout his life: that Jesus is fully God and man, and that a heart-directed relationship with God is available for everyone. Jesus' birth in a rented stable is not what might be expected for God's own Son and yet angels authenticate the baby as Christ, God's chosen one. Jesus' birth demonstrates that God is accessible and unashamed to identify with the lowly. Jesus ushers in a new type of relationship that he refers to as the kingdom of God having come among the people.

Relatively little is recorded of Jesus' early years. The Gospels focus on the period beginning with Jesus' baptism and ending with his death three years later. As he emerges from the baptismal water, the heavens open and God's voice declares: "You are my Son, whom I love; with you I am well pleased."[1] Following this public authentication of Jesus' divinity, he departs to spend a solitary time in the wilderness, fasting and praying in preparation for his public ministry. Physically weakened, Jesus is unsuccessfully tempted by the devil to do three things: to provide food by using his miraculous powers, to put himself in physical danger so that he will be rescued by God's angels, and to worship the devil in order to lead the world. Jesus responds with biblical quotations that show how his relationship with God is his first priority, and that God will provide for those who seek him.

Jesus began his public ministry by choosing twelve close disciples. During the three years that the Gospels describe, Jesus lived with these disciples, teaching, healing, and doing miracles as he traveled in and around Galilee and on the eastern side of the Jordan. Finally, Jesus returned to Jerusalem in triumph on what is now known as Palm Sunday because the crowds placed palm branches on the road in deference to Jesus as a spiritual leader. The triumph subsequently turned to grief as Jesus was arrested, crucified, and buried. The story does not end there, however, as these same disciples witnessed him in bodily form, a resurrected Jesus, three days later.

The Gospels describe Jesus as a man whose actions and teaching come from a deep understanding of God's will. When challenged on his teaching, Jesus quoted the Bible and pointed to his miracles as proof of his divinity. Only in a few instances did Jesus directly self-identify as God's Son,[2] preferring instead to let his teaching and miracles speak for themselves. In general, Jesus uses the ambiguous term the "Son of Man." On the surface, the title seems to stress his humanity, but with biblical

1. Mark 1:11
2. Matt 16:13–20

overtones that mean that this title is best translated as "exalted one." Jesus was consistently subtle in revealing his identity and his message, which was clear to those with open hearts and minds.

Roughly one third of Jesus' recorded teachings are in the form of parables: pithy, memorable stories that powerfully illustrate instructive truths or principles, by drawing the listener in to identify with the people or situations they depict. Through his teaching, Jesus brought a new understanding of God's desire for a relationship with each individual. Jesus' teaching scandalized the religious authorities, who considered themselves essential mediators to God. One of Jesus' best known parables is the story of the "good Samaritan" which illustrates how an individual's relationship with God should direct their relationship with others:

> On one occasion an expert in the law stood up to test Jesus. "Teacher," he asked, "what must I do to inherit eternal life?" "What is written in the Law?" he replied. "How do you read it?" [The man] answered, "'Love the Lord your God with all your heart and with all your soul and with all your strength and with all your mind'; and, 'Love your neighbor as yourself.'" "You have answered correctly," Jesus replied. "Do this and you will live." But [the man] wanted to justify himself, so he asked Jesus, "And who is my neighbor?"
>
> In reply Jesus said: "A man was going down from Jerusalem to Jericho, when he was attacked by robbers. They stripped him of his clothes, beat him and went away, leaving him half dead. A priest happened to be going down the same road, and when he saw the man, he passed by on the other side. So too, a Levite, when he came to the place and saw him, passed by on the other side. But a Samaritan, as he traveled, came where the man was; and when he saw him, he took pity on him. He went to him and bandaged his wounds, pouring on oil and wine. Then he put the man on his own donkey, brought him to an inn and took care of him. The next day he took out two denarii and gave them to the innkeeper. 'Look after him,' he said, 'and when I return, I will reimburse you for any extra expense you may have.'
>
> "Which of these three do you think was a neighbor to the man who fell into the hands of robbers?" The expert in the law replied, "The one who had mercy on him." Jesus told him, "Go and do likewise."[3]

---

3. Luke 10:25–35

The parable of the Good Samaritan is typical of the stories Jesus used to convey the truth and goodness expected from a true follower of God. The story captures the attitude expected in helping others and, because of the antagonism between Jews and Samaritans, models how true love and grace transcends religion and culture. The parables succinctly convey profound truths about people's character that encourage listeners to adopt the same values. Jesus' parables speak to the heart; part of the reason why Jesus is widely considered to be a great teacher.

Perhaps nowhere is Jesus' character more deeply revealed than in the Sermon on the Mount.[4] The sermon begins with a list of who is blessed: those who are poor in spirit, the meek, the peacemakers, the pure in heart, and those who thirst after righteousness. The list is not an exhaustive set of requirements for living a spirit-directed life, but illustrates the values and characteristics of those who earnestly seek after God. Jesus taught that the good life is a life lived in seeking right relationships. His aim was not to impart information but to change lives.

The Sermon on the Mount captures how the law guides behavior by instilling an internal understanding of the spirit of the law. Jesus takes the commandments against murder, adultery, and stealing, and extends the principles from outward actions to internal motives. The commandment "Do not murder"[5] changes from being an issue of restraint to a matter of the heart: "anyone who is angry with his brother will be subject to judgment."[6] The prohibition against adultery is extended to lust, which Jesus called committing adultery in the heart.

Jesus' teaching restored the full understanding of the Mosaic Law and emphasized how following God is more a matter of inner character than following rules. At the core of Jesus' teaching is an understanding that embracing true religion guides a person's thoughts which in turn directs behavior. Right living is more than following laws but is a continual, transformational development of the inner life. Jesus taught that religion is not rules but relationship. The incarnation of Jesus, as God-made-man, is the model of the relationship intended between God and man. Jesus acts with a confidence in God's provision that forms the core of his being and guides him through all situations no matter how challenging.

---

4. Matt 5–7
5. Deut 5:17
6. Matt 5:22

PRAYER

Prayer is a central component of Jesus' life. Jesus began his public ministry with an extended time of prayer in the wilderness and concluded his life in prayer to God during an agonizing death by crucifixion. Throughout his life, Jesus publicly and privately modeled the prayerful relationship with God that is intended for each person. Jesus prays before all the main events of life: the start of his public ministry, the choice of his twelve close disciples, and his final entry to Jerusalem. Jesus teaches his disciples to pray in private, to pray thoughtfully, and to pray from the heart.

When asked how to pray, Jesus provided a model that acknowledges God and then asks for physical and spiritual sustaining.

> Our Father in heaven, help us to honor your name. Come and set up your kingdom, so that everyone on earth will obey you, as you are obeyed in heaven. Give us our food for today. Forgive us for doing wrong, as we forgive others. Keep us from being tempted and protect us from evil.[7]

The prayer Jesus taught the disciples provides a model for a progression between different forms of prayer. Beginning with a contemplation of God's presence, the prayer moves on to intercessory prayer, which asks God to act in each person's life and the lives of others. Prayer includes a broad array of activities, but at the core, prayer is communion with God. Prayer can take several forms, including experiences in organized worship, personal and petitionary prayer, praise, meditation in stillness, and simply a contemplation of God in nature.

Prayer, particularly petitionary prayer, assumes that God can and will act in the world to bring about change. Jesus invited exactly this type of prayer:

> Ask and it will be given to you; seek and you will find; knock and the door will be opened to you. For everyone who asks receives; those who seek find; and to those who knock, the door will be opened.[8]

In this and many other passages, prayer is the vehicle for a divine relationship with each person. Prayer is the system of communication that God wants to establish to bring about change in the world, beginning with personal spiritual change. God's action in the world is intrinsically

---

7. Matt 6:9–13 (CEV)
8. Matt 7:7

linked to an individual's expectation that God will act. Jesus often asked people what they wanted, miraculously answering prayer in ways that correlate with a person's faith.

> "What do you want me to do for you?" Jesus asked him. The blind man said, "Rabbi, I want to see." "Go," said Jesus, "your faith has healed you." Immediately he received his sight and followed Jesus along the road.[9]

Prayer is not a mechanical operation in which a petition will lead to a predictable outcome. In the words of Thomas Aquinas, one purpose of prayer is "to remind ourselves that in these matters we need divine assistance."[10] Understanding that prayer is not just petitioning God for provision or intervention but a personal encounter with God to aid spiritual development explains why God does not always act to optimize short-term happiness. Prayer may appear to be one-sided with God's response difficult to hear or understand. Unanswered prayer does not mean divine inattention but may be a 'not now' or 'not yet' refocusing an immediate desire toward deeper spiritual growth. Even Jesus experienced unanswered prayer in his cry from the cross: "My God, my God, why have you forsaken me?"[11]

Prayer is the personal meeting with God; a divine interaction that is relational and unscripted. Prayer fosters a sensitivity to God's direction for each person's life. Millions of people over the centuries have claimed that answers to prayer provide personal proof of God's existence. As a personal experience, answers to prayer are necessarily subjective and must be interpreted within a uniquely personal context.

The goal of much prayer is an alignment of individual and divine wills. Prayer and God's will can be imagined as two waves moving between God and an individual. If the prayer wave aligns with God's will, the two waves will be in phase, amplifying the wave the same way that powerful laser light results through a phase alignment of many individual waves. In the spiritual realm, an in-phase prayer leads to an increased likelihood of the prayer being answered. Waves of the same amplitude with different phases interact destructively to give no signal. Corporate prayer, in which many people pray together, functions differently from individual prayer, not because more pressure is placed on God, but rather

9. Mark 10:51
10 Quoted in Polkinghorne, *Science and Providence*.
11. Matt 27:46

because, with more people, there is a greater chance of an alignment of wills. Even if only a small minority find their prayers align with a divine will, the amplification can be sufficient to direct the ultimate outcome. Corporate prayer creates a setting with greater sensitivity to the perspectives of others, including God, to help achieve a consensus that aligns with God's will. In this way, prayer is neither mechanical nor magical but rather is a personal encounter between God and each person.

MIRACLES

God's interaction in the world is sometimes separated into general or special providence. General providence is the divine sustaining of the world's regular processes that emanates from God's faithfulness, described scientifically as laws of nature. God's direct action, or special providence results in miraculous physical changes. Examples of special providence include the plague of locusts sent by God to persuade Ramses II to let the Israelites leave Egypt and Jesus' healing miracles.

Miracles feature prominently in the ministry of Jesus. Jesus' miracles, most of which involve healing, authenticate his teaching. When John the Baptist wonders if Jesus is the promised Messiah, Jesus sends him the message: "The blind receive sight, the lame walk, those who have leprosy are cleansed, the deaf hear, the dead are raised, and the good news is proclaimed to the poor."[12] Although John is not explicitly answered, the response invites him to interpret these miracles as a sign of Jesus' divinity.

The popular conception of a miracle is an extraordinary event that goes beyond normal experience. A theological definition adds that the event is caused by God and has the spiritual intent to increase faith. The spiritual significance conveyed by miracles requires personal interpretation; miracles are not conclusive proof of God's existence. Despite Jesus' diverse healing miracles, mass provision of food after long sermons, and walking on water, his three dozen recorded miracles failed to convince most of the religious authorities of his divinity.

Behind both definitions of miracles is a belief in an orderly universe where the laws of nature are constant and expected to continue in the same pattern forever. The biblical writers attribute both the mechanisms of natural events and miracles to the activity of God. The biblical focus in each case is on God's providential care through either ordinary recurring

---

12. Matt 11:4–5

events or through miracles. From the theistic viewpoint, the regularity and predictability of the created world stems from the faithfulness of God.

For religious people, while miracles lie outside the bounds of normal physical experience, they are consistent with the nature and expression of God's activity in the world. Far from breaking normal patterns, miracles depend on those patterns in two ways. First, without regularity and consistency in nature there cannot be a remarkably significant exception. The complexity of the natural world is understandable precisely because the regularity allows scientific instruments to detect patterns and forces from which fundamental knowledge is attained and able to be predicted in the future. Second, after the exceptional event occurs, the resulting situation returns to familiar, predictable patterns.

There are two broad types of miracle. The first type is an arranged coincidence where natural processes occur with an exquisite sense of timing to endow the situation with religious meaning. A classic biblical example is when the apostle Peter is asked to give the priests a temple tax for himself and Jesus. Jesus instructs Peter, a fisherman, to "go to the lake and throw out your line. Take the first fish you catch; open its mouth and you will find a four-drachma coin. Take it and give it to them for my tax and yours."[13] This is not outside the boundaries of nature. Fish are often attracted to shiny objects, and are occasionally found to have swallowed coins. The divine coincidence here is two-fold. First, the timing—for Peter to catch a fish containing a coin at exactly the point told in advance by Jesus, and secondly the coin's value being precisely that required for the tax.

Many people experience similar coincidences. Although coincidences may not be miraculous, the events may be interpreted as carrying divine meaning. Meeting someone who relays particularly valuable or timely information, or discovering an item or idea that proves particularly influential could well be understood as miraculous. Coincidences which appear to be chance to an outside observer may be interpreted as providential or even as miraculous by the individual depending on the personal relevance, significance, and context. People interpret such events as chance, fortuitousness, or providence, descriptors that increasingly acknowledge a divine role in the process.

13. Matt 17:24

The second type of miracle is an extraordinary event that does not fit within the expected pattern of normal experience. There is a long tradition of viewing miracles as accelerated natural processes, which avoids the impression of God working against natural forces. Augustine used the idea of an accelerated fermentation to explain Jesus' first miracle of turning water into wine. A similar acceleration was often used to explain miraculous healings. Accelerated natural processes, however appealing to great minds of antiquity, are revealed by scientific advances at the molecular level to be inherently unnatural. Physicist-theologians have sought a deeper understanding of divine involvement to explain events outside normal experience.

Science and religion focus on two different aspects of miraculous events. A miracle is a unique, nonrecurring event filled with divine meaning that stands apart from the regular, observable events studied by science. Scientific laws typically rest on the orderliness and regularity of natural processes, which allow accurate predictions of what will probably happen in the future. Miracles are inherently non-reproducible. Science cannot pronounce a given extraordinary event, a miracle, to be impossible, only that the event is foreign to previous experience and highly improbable on the basis of empirically derived theories. Scientific laws and theories do not prescribe or legislate what must happen but rather describe or explain experiences based on repeated events or observations. Science does not preclude the occurrence of a miracle. Indeed, many scientific theories incorporate non-reproducible events within a continually evolving understanding of the way the world is.

Unique, nonrecurring events are an essential part of some sciences. In astronomy, supernova explosions are recorded, analyzed, and interpreted. Particle physics has benefited immensely from "golden events" in which the movement of fundamental particles reveals truths about the structure of the atom. Earnest Rutherford fired alpha particles at gold foil and found that while most particles pass through, a few were violently deflected. Those few observations led to a new model of the atom in which most of the atomic mass resides in the dense nuclear core. Rutherford wrote that the event "was almost as incredible as if you fired a 15-inch shell at a piece of tissue paper and it came back and hit you."[14] Extremely rare events, perhaps occurring only one time in a million, can

---

14. Andrade, *Rutherford and the Nature of the Atom*, 111.

provide true insight into the nature of reality that compliments information from reproducible events.

How an intangible God answers prayer or causes miracles by influencing either intangible thoughts or physical processes is unknown. The regularity of nature allows life to flourish and yet appears sufficiently malleable for God to act within the world. No mechanism or "causal joint" has been identified at which a divine influence acts, assuming that God's agency would sufficiently stand out from the regularity of nature to be detected. The same type of search for individual agency is underway in neuroscience to understand how a person's mind is able to trigger bodily movement.

THE CAUSAL JOINT

If God is responsible for the entire existence of all that is, and continues to be involved in people's lives, there must be a nexus point—a causal joint—between the physical and divine realms. If God answers prayer, performs miracles, and influences people's lives then shouldn't divine involvement be detectable? The answer, strangely enough, is no, not always.

In the early twentieth century several theologians developed a new approach to explain divine activity. This new system of thought, known as process theology, focused on God as a persuader who must work within the rules of the process, a fellow-sufferer who understands the human condition who participates in the events and is changed by them. In process theology, individual entities have freedom to ignore the divine persuasion, absolving God of a responsibility for moral and natural evil. God chooses to limit involvement in the world, or is only capable of limited influence. God acts through the interior nature of each organism through a kind of divine potential existing in all of nature. A consequence of process theology is that while God intends the best for creation, divine persuasion is inadequate to prevent wars, genocide, and natural disasters.

Critics of process theology argue that in this theory God is reduced to an influence that is not consistent with the traditional understanding of an omnipotent being. Rather than a sovereign God who performs miracles to release the Israelites from Egypt, heal the sick, and raise Jesus from the dead, the God of process theologians lurks in the indeterminate world carefully calculating probabilities to avoid detection. Orthodox Christian theology is built on the belief that ultimately God will redeem the world, purge evil, and establish a heavenly kingdom. Process theism can have no such assurance about the final outcome.

An alternative location for God's purposive action lies in the quantum realm where events are inherently unpredictable but have a statistical regularity. The approach provides a locus for divine activity that evades detection while allowing acts that are the result of activity characteristic of a divine, omnipotent agent. The unpredictability inherent in nature is captured in the Heisenberg uncertainty principle that the momentum becomes less precise as the position of a particle becomes better defined, and is not an inability of science to understand or describe natural processes. An understanding of the universe derived from physical observation leads to an incomplete description of reality. Consequently there exists a space where a divine influence might remain beyond the limits of detection, hidden in the spurious chatter of statistical noise. The incomplete rules of nature leave room and flexibility for minute changes at the atomic level that could potentially accumulate to cause changes at the physical level of everyday experience. A divine nudge in the quantum realm then influences events, resulting in a change in some process at the human scale that has religious significance. What appears to be a chance event may be caused by divine influence. God need not direct an electron in a specific way but rather is able to actualize one of several pre-existing possibilities.

A different explanation of divine activity imagines that God's influence is an emergent property of reality. God acts by exerting pressure akin to the way in which people put thoughts into action. The mechanism describing how this might occur is obscure because at the current time there exists only a limited understanding about how mental thoughts influence responses in the human body. Theologians speculate that the causal joint may be through the transfer of information rather than energy. Exertion of a divine influence could influence individuals and physical processes. God would simply have to introduce the information to guide processes through to a divine ideal. There is considerable speculation on whether information can be imparted without an obligatory energy transfer, a process that would be experimentally detectable.

Scientist-theologian John Polkinghorne prefers to locate God's activity in complex dynamical systems, the realm of modern chaos theory. Chaos theory is a relatively recently discovered phenomenon where very small changes in the initial conditions are amplified to cause profound changes in dynamical systems, essentially reclaiming an element of indeterminacy from even Newtonian systems. Chaos theory's essence is illustrated in the butterfly effect; the beating of a butterfly's wings in Hawaii causes an air disturbance that is amplified in air currents to ultimately influence the weather in New York. The collisions of molecules in a gas behave somewhat like small colliding billiard balls that are so strongly influenced by remote events that even the variation in the gravitational field due to an extra electron on the other side of the universe influences the outcome of apparently random motion. God's action in the system is veiled in a cloudy unpredictability that is none-the-less discernable by faithful believers.

Critics point out that the equations describing chaos theory are deterministic. In other words, there is no openness of nature to divine influence because all the infinitesimal influences are determined but are too complex to be tracked down. Whether or not chaotic systems provide a viable causal joint is an open question. Even if chaos theory does operate as a divine conduit, this theory still offers a very limited range of divine influence outside the boundaries of scientific detection. These are only a few of the potential places at which divine influence could cause dramatic physical changes at the scale of human experience.

The very nature of divine activity, as described in the Bible, means that God both permanently upholds the predictable nature of the world and acts outside that predictable nature as an answer to prayer and providence. An inextricable entanglement exists between these two roles. Under the same divine guidance that ordinarily leads nature to behave predictably, nature may sometimes be induced to behave differently from normal through a discontinuous change that is in character with nature but very infrequently accessed. As an analogy, water behaves normally until 100 °C at which point water suddenly becomes steam. Water has a dramatically altered physical state at the boiling point, though the underlying laws of physics do not change.

A religious equivalent of this discontinuity might be the way in which prayer for healing is sometimes answered. Occasionally, examples surface of believers who have medical examinations showing the presence of a fatal tumor that, after much prayer, is subsequently unable to be detected. Are the body's repair mechanisms able to be harnessed through divine fiat in ways that still cohere with the essential character of biochemical processes?

All of these questions are important, but they do not touch the underlying experience of the miraculous. Religious *experience* is not concerned with the mechanism of God's involvement but the interpretation of the events as evidence of God's continuing interaction with people in and through nature. Science and religion provide insight into the causal joint, but their approaches differ because of the framework from which each seeks an explanation. Science seeks to understand miracles through information obtained from the physical universe, whereas religion begins with the assumption that God's nature transcends understanding to the extent that some of God's actions may be beyond people's understanding. A helpful analogy is to imagine a game developed by a computer programmer. If the creator is good, then a new game will have a coherent sequence of events. An observer will identify consistent patterns from which the game's object can be understood and the laws of motion determined. A keen gamer might even develop a set of rules for the game, provided the creator remains faithful to the unfolding spirit of the events. If a previously unobserved play occurred, this would be duly noted in the expectation that ultimately the observation would be understood.

The illustration provides a helpful image for understanding God's role in creation. Miraculous events bear the consistency of God's character in the same way as his faithfulness sustains the world's regularity to allow life to flourish. An important difference between the game analogy and the religious understanding of God's involvement in the world is that only in the biblical perspective do the characters know their creator. Miracles are a reminder to religious individuals that God remains alive and active in his world.

The world people experience has an openness or flexibility that allows for personal choice. Rare instances where the hand of God does intervene serve to highlight his role as a constant, invisible, transcendent presence. An advantage of the quantum and chaotic models of the causal joint lie in a continuous divine presence within the existing structures of the world. The non-interventionist goal is to allow divine action without violating the laws of nature while evading detection by scientific methods. The continuous divine engagement pursued in these different approaches fits the biblical theme declared by the apostle Paul that God's presence is evident from creation, not from inexplicable oddities in nature. The Bible clearly provides a picture of God interacting with his people, suffering when they suffer, rejoicing in righteousness, and delighting in grace, and doing this within the bounds of the natural world. Prayer and miracles reinforce the relationship between God and individuals and the central biblical message that the kingdom of God is accessible to everyone.

THE RESURRECTION

The resurrection is the ultimate miracle supporting Jesus' claims about himself and his divinity. Evidence for Jesus' death, resurrection, and subsequent appearances appears primarily in the four Gospels in the New Testament, though this is verified by independent historical writings of the time. Some differences exist between the four Gospel accounts, reflecting the style of writers two thousand years ago who were concerned about capturing the key essence of people's lives and were less concerned with precise details.

One reason to believe that the Gospels are an accurate record of what the writers witnessed is that the accounts include events that the writers would have found disturbing. For example, the Bible records that Peter, the early church's leader, three times denied that he even knew Jesus. Moreover, the Gospels describe Jesus' first appearance being to several women. Women were considered unreliable witnesses at the time the Gospels were written, making the inclusion of the women's testimony better fit an accurate account of events than a fabricated story. Occasional differences between the Gospel writers' accounts of Jesus' death likewise point to their accuracy—different observers naturally see events from their own, differing points of view. The Gospels' inclusion of the disciples' foibles, several unsettling teachings, and Jesus' miracles, make more sense in a truthful account of Jesus' life than in a revisionist history where such details would be expected to be removed. While mythological stories and histories of the time did include miraculous events, these were not believed by many of the common people or the skeptical Roman and Greek audience for which biblical writers wrote.

By far the largest part of the Gospels is devoted to describing events leading up to Jesus' crucifixion, death, and resurrection. Because of the significance of the resurrection, numerous analyses of Jesus' death and resurrection have appeared. After the crucifixion, Roman soldiers, well-acquainted with death, plunged a spear into Jesus' side to ensure that he really was dead before allowing his followers to take the body. Jesus' body was prepared for burial according to Jewish custom, placed in an unused tomb, and the tomb opening secured with a large boulder. The Roman governor ordered soldiers to guard the tomb to ensure that Jesus' disciples did not steal the body. At dawn, a violent earthquake caused the stone to roll away and an angel greeted the women who came to the tomb with the news that Jesus was alive. In an ironic twist, the soldiers guarding the grave became witnesses of an empty tomb that they were sent to ensure never came to exist.

Historical documents and projected reconstructions of people's actions, support Jesus' resurrection even though the event transcends normal experience and occurred two thousand years ago. Religious and Roman authorities of the time needed only to produce Jesus' body to quash any rumors of resurrection. Not only were the authorities unable to provide Jesus' body, but several witnesses claimed to have seen the risen Jesus. The apostle Paul lists appearances of the resurrected Jesus to several hundred people over a period of forty days. Jesus' first-century disciples were convinced of his resurrection and spent the remainder of their lives teaching the gospel message and proclaiming Jesus' resurrection as proof of his divinity. Ten of the eleven remaining disciples suffered martyrdom rather than repudiate their convictions and the eleventh disciple, John, died in exile.

After Jesus' resurrection, the early Christian community adopted Sunday as the day of worship because this was the day on which Jesus had conquered death. Adoption of the new worship day is congruent with a widespread belief in the resurrection.

The apostle Peter's first recorded sermon describes the resurrection of Christ as inevitable. God raises and delivers Jesus from death "because it was not possible for Him to be held by it."[15] Jesus' nature as fully God and fully man is congruent with Jesus' body suffering death and with death being impermanent because God could not be destroyed in his own creation.

Christian theology understands Jesus' resurrection as the prototype for his followers. Resurrection into a new type of body is required to escape the endless cycle of death that necessarily comes from being Homo sapiens. God's final purpose in offering to share eternal life with each person therefore requires a second creation event where death is banished for those who have chosen to seek him. The precise nature of the second creation is withheld, but Jesus' resurrection is a foretaste of what awaits his followers. Although the specifics are not provided, the requirements for entry are clear: only those who acknowledge Jesus as Lord and repent of their sinful nature are to enjoy eternal life with God.

According to Christian belief, Jesus' resurrection is the prototype for each person to follow at their own appointed time. Jesus' post-crucifixion

---

15. Acts 2:24

appearance to the disciples in a locked room demonstrates that his resurrected body is fundamentally different in being able to pass through solid objects. Jesus' body was both recognizable by the disciples and yet clearly different. Jesus ate and offered to let doubting Thomas place his hand on his body where he was crucified, demonstrating the tangibility of Jesus' resurrected body.

Each person's attitude toward Jesus' resurrection depends upon what they believe happens at death. A Christian's hope of resurrection is based on the fact of Christ's own resurrection. Resurrection requires continuity with the old body and a person's unique spiritual and physical identity, but in an immortal form unlike that of the original body. The form of these resurrected bodies must necessarily be physically different from earthly bodies because a perfect relationship with a good God requires an immortal body devoid of decay and the sinful desires inherent in the biological history of the human body. The first body is the old creation, one with the potential to influence the ultimate form embodied in the resurrected body, the new creation.

Bodily resurrection is foreign to normal experience. Discussions of life after death are therefore speculative. One appealing model of the resurrection is that a person's unique individuality survives in the mind of God and can be recreated by God at the resurrection. The individual atoms composing the body are less important than the information that specifies an individual person, their humanity, their relation to others, and, most importantly, their relation to God. At death the pattern is dissolved but able to be reconstituted by God at the time of the resurrection. Support for this perspective is the perpetual replacement of the individual atoms in the human body. Eating provides proteins, minerals, and other nutrients that are broken down into constituent molecules or even specific atoms, that are then incorporated into the body. Over the course of roughly seven years the atoms in the body are replaced and yet individuals remain the same *person*.

Jesus' resurrection is the ultimate endorsement of his teaching that true life involves a right relationship with God and a right relationship with others. Right relationship is at the core of the Sermon on the Mount which describes life characterized by love, justice, and mercy. Life's true fulfillment will not be achieved through either physical pleasure or a slavish adherence to religious performance. Ultimately, resurrection to a new and perfect life is required to overcome the injustices inherent in a world in which evil exists and to establish permanent justice.

CONCLUSION

Jesus is, at the same time, the most attractive and the most divisive figure in human history. Jesus' message of love, justice, and mercy has universal appeal, but he requires a personal allegiance that runs counter to natural human tendencies. Jesus' good news is that the kingdom of God is available to all who follow God's will rather than their own desires. His life was a demonstration of obedience to that will, even to death.

Central to Jesus' life was a vibrant practice of prayer. Jesus' prayers for his disciples and for himself provide a model for his followers. Jesus' many miracles, and particularly the healings, show how prayer for physical transformation is intimately linked to spiritual development. Jesus demonstrates that prayer, direct communication with God, has the power to change the world. Jesus' prayer for his disciples to spread God's message ultimately transformed the Roman world and continues to work through human hearts today.

Prayer and miracles work differently from the usual physical laws of cause and effect. Prayer changes and transforms lives, and may lead to tangible changes in the world. Scientific analysis is the wrong tool to track God's response to petitionary prayer because while science does not preclude miraculous events, science generally rests on reproducible events. These align with the faithful nature of God's character, manifest in the regularity of nature. God's primary concern in both prayer and miracles is spiritual development and personal relationship with individuals.

Jesus' greatest miracle is the resurrection. Jesus' death and resurrection lie at the core of Christian belief because a bodily resurrection confirms the hope of eternal life. Life after death is foreign to both everyday experience and science. Strong circumstantial evidence points to the veracity of Jesus' resurrection, though the mechanism of divine action is not revealed. The marks of the crucifixion on Jesus' body suggest that individual identity will be preserved in the resurrected life, consistent with his message that relationship with God and others is key to understanding life.

The resurrection lies at the core of Christian belief. Over the centuries there have been numerous explanations for, and arguments against, Jesus' bodily resurrection. Jesus revealed himself to be God's Son in a way that invited belief because faith is an integral part of the gospel message. The evidence for Jesus' divinity and God's desire for a relationship with each individual person is subtle. God does not provide evidence that compels belief in his existence but appears to have left ambiguous hints that are open to interpretation. God seems to require a level of personal engagement that, with faith, leads to personal conviction. God invites the individual to interpret the events as evidence for his existence. There is no celestial website where God answers questions. Only through personal belief can the ultimate experiment be run to test God's existence, one human heart at a time. God is neither completely hidden nor unequivocally present, but reveals himself to those who seek him.

DISCUSSION QUESTIONS

1. In John's Gospel, chapter 10, Jesus claims to be the only way by which a person can know God: "I am the gate; whoever enters through me will be saved."[16] What are the most compelling reasons for and against Jesus' claim?

2. Science provides powerful explanations in realms that extend from physics to social science. Has science removed the need for the existence of God?

3. Many people attend church on Christmas and Easter who are not Christian. Why is this?

4. Does scientific knowledge make believing in Jesus easier or more difficult?

5. Prayer is the most personal element in an individual's relationship with God. Given that petitionary prayer is personal and open to interpretation, can the veracity of an individual answer to prayer be taken as evidence for God's existence?

6. Are miracles best thought of as stopping the normal laws of the universe, superseding the laws of the universe, an injection of God's will into the universe, or something else?

7. Miracles have an essential religious component that encourages belief. What type of miracle would you find most compelling and why?

8. Presumably God could have stamped each person with some type of divine label pointing to his existence, a scientifically identifiable equivalent of "Made in Heaven." Why did God prefer the evidence of the resurrection instead?

16. John 10:9

9. Traditional Christian theology holds that Jesus Christ was simultaneously God and man. Is this article of faith necessary for someone to consider him/her-self a Christian?

Further reading for "Jesus Christ: Prayer, Miracles, the Causal Joint, and the Resurrection"

1. John Polkingnorne, *Science and Providence: God's Interaction with the World.* Conshohocken, PA: Templeton Foundation, 2011. John Polkinghorne, a trained physicist and theologian, brings refreshing insight into many perplexing theological issues: miracles, prayer, and evil. The focus of many of Polkinghorne's ideas is on how God might interact in the world in ways that harmonize with orthodox Christianity and advances made in quantum physics on the nature of matter.

2. John Polkinghorne, *Exploring Reality: The Intertwining of Science and Religion.* New Haven: Yale University Press, 2007. Polkinghorne compares scientific and religious perspectives on several difficult issues at the intersection of the two disciplines. Examining the limits of an empirical approach to reality, Polkinghorne argues that human experience is fully manifest only through a religious understanding of reality.

3. Malcolm A. Jeeves and R. J. Berry, *Science, Life, and Christian Belief: A Survey of Contemporary Issues.* Grand Rapids: Baker, 1999. The authors focus on five main topics: origins, ontology, biology, psychology, and ecology. Covering topics ranging from biological evolution to the nature of the soul, the authors provide insight from modern science to bear on perplexing questions of how religious belief can be compatible with scientific discoveries.

4. Craig L. Blomberg, *The Historical Reliability of the Gospels.* Downer's Grove, IL: IVP, 1987. Blomberg sketches out the modern approaches taken to understand the Gospels before examining differences in the Gospel accounts. Blomberg's approach to understanding miracles and the resurrection lies in understanding science, the fulfillment of prophecy, and understanding Jesus' role as God's son.

5. Charles Hummel, *The Galileo Connection.* Downers Grove, IL: IVP, 1986. A widely read and well-respected introduction to "resolving

conflicts between science and the Bible" as promised on the front cover.

6. Dallas Willard, *The Divine Conspiracy: Rediscovering Our Hidden Life in God*. New York: Harper, 1998. Dallas Willard's conspiracy is uncovering God's message for mankind through an exposition of the Sermon on the Mount. The beauty of the book lies in focusing on human will and the vagaries of living a virtuous, religious life in accordance with God's will.

7. Philip Yancey, *The Jesus I Never Knew*. Grand Rapids: Zondervan, 1995. Yancey tries to imagine himself living alongside Jesus and responding to Jesus' teaching. In blatantly honest prose, Yancey evaluates his thoughts and actions on the new standard that Jesus taught and lived. The book is not for the faint of heart but for those truly wanting to change themselves to follow God more closely.

8. Frank Morison, *Who Moved the Stone?* Grand Rapids: Zondervan, 1987. The classic popular analysis of Jesus' resurrection. Frank Morison was an English journalist who intended to debunk Jesus' resurrection by studying the available evidence only to end up believing that the resurrection was a real event.

9. N. T. Wright, *The Challenge of Jesus: Rediscovering Who Jesus Was & Is*. Downer's Grove, IL: IVP, 1999. Bishop Wright has devoted much of his scholarly writing to address Jesus' character and resurrection. The book represents an accessible overview of Jesus that draws from Wright's extensive research and defense of the resurrection.

6. A Brief History of Science: From Prehistory to Particle Science

SCIENCE HAS BEEN NURTURED in the cradle of religion from the earliest human history. From prehistoric times, people have exhibited an innate yearning to understand how the world came to be. Unusual natural events such as lightening, thunder, and earthquakes were frequently explained as divine activity by many cultures. Priests were engaged to interpret these natural signs and to act as mediators to secure the favor of the gods. Over time the priestly interpretation of divine messages coded in natural events changed into a logical analysis of patterns based on repeated observation. This transition loosely correlates with the early beginnings of science.

Many early civilizations believed that gods existed in celestial bodies. Priests in these civilizations made very accurate observations of stars and planets in order to correctly plan feast days, to interpret what the gods wanted, and to understand the likely influence on future events. Astronomical observations provided an understanding of planetary motion that made scientific analysis a priestly duty. The result is a legacy of very accurate records of celestial events spanning several millennia.

As civilizations rose and fell, different cultural and religious influences changed the direction of scientific inquiry. Of all the early cultures, Greek thought exerted the most profound influence on the development of science. Greek thought was subsequently viewed as definitive by the Romans who compiled summaries of Greek writings rather than search for new ideas. The secular Grecian ideas, particularly those of Aristotle, were Christianized over many centuries and were largely unchanged until Galileo's experimental method demonstrated that many Aristotelian ideas were wrong. Galileo's observations simultaneously demolished the Aristotelian framework based on causes, and established a type of analysis that marks the beginnings of modern science.

Science has advanced less from steady incremental progress, moving from simple particles to particle physics, than through a haphazard series of failures and successes. Continued refinement has led to improved ways to gain knowledge from the natural world. The inherent self-correcting nature of scientific inquiry has benefited science and religion, and while the interactions have been profound, science is generally viewed as the surer path to understanding. An historical examination of some of the influential civilizations and figures in science provides valuable guidance for understanding the way science and religion interact and for avoiding the kinds of philosophical conflicts that have arisen in the past.

EGYPTIAN SCIENCE

The Fertile Crescent, encompassing semi-arid Western Asia, the Nile Valley, and the Nile Delta of northeast Africa, formed a crucible from which many early civilizations emerged. Deserts surrounding the Nile River isolated communities from outside attack while the predictable river floods brought an annual renewal of rich topsoil that allowed for a plentiful food supply. Between the sixteenth and the eleventh centuries BC, the Egyptians were at the pinnacle of their culture and developing an empire. The most famous legacy of the Egyptians—the pyramids—illustrates their unified understanding of science and religion.

The pyramids were built on the western side of the Nile where the sun god Atum Ra appeared to die each night. Ancient Egyptians believed that in order for souls to reach the afterlife, they needed to follow the path of the dying sun god. During the journey the body needed to be preserved to provide a home for the soul. The pyramids, which were built both to aid the dead to reach the afterlife and to demonstrate the power of the Pharaoh, stimulated the development of considerable technical knowledge.

During construction of the pyramids, stones weighing up to three tons were cut, moved, and positioned with great accuracy using only levers and ropes. The expertise required for this unparalleled building project appears to have been gained through a repeated give and take based on experience rather than from the development and application of first principles. The approach is illustrated in Egyptian mathematics, where the formula for calculating the area of a circle of diameter d is Area = $(d - d/9)^2$ was likely determined by trial and error. Rearranging the equation gives a value for π of 3.16 which is very close to the 3.14 determined by the Greeks.

One of the Egyptians' most important contributions to science was the development of the solar calendar. Religious motives were at the heart of the quest to calibrate time for accurately predicting the day and hour of feasts and festivals. The Egyptians developed the sun dial for measuring time during the day and water clocks for the night. Egyptians refined the lunar calendar by correlating the year with the rise of the star Sirius, eventually leading to a calendar year of twelve months of thirty days with five feast days.

BABYLONIAN SCIENCE

The Babylonian culture, located in modern-day Iraq, reached a golden age in 1800 BC under King Hammurabi, whose name is usually associated with a code of laws concerning property rights, business, family, labor, and injuries. The Babylonians viewed the five visible planets and the sun and moon as gods, keeping careful records of the motions of stars and planets to understand what the gods were doing. Celestial observations were made by priest-astronomers from terraced brick towers, ziggurats, which served as both religious temples and viewing platforms. Interpreting the movements of the celestial bodies allowed the priests to use planetary motion to create a calendar of religious observances. Over time, the Babylonians grouped the stars into constellations named after various animals that formed an astronomical zoo. The Babylonian observations, and some of their mathematical work, was subsequently passed on to the Greeks and became the basis for many of their astronomical generalizations.

During the New Babylonian Empire, 625–539 BC, planetary motions were used to develop a personalized form of astrology. According to this system, each of the deities influenced one day of the week, a belief that was ultimately incorporated into the modern week-day names, in English: Sunday for the sun, Monday for the moon, Saturday for Saturn, and, in French Tuesday/Mardi for Mars, Wednesday/Mercredi for Mercury, Thursday/Jeudi for Jupiter, and Friday/Verdi for Venus.

GREEK SCIENCE

The Greeks were the original scientists of ancient Europe. They believed that rational enquiry was the only prerequisite for understanding the world and studied the world in order to make sense out of nature. Unlike previous civilizations, the Greeks pursued knowledge as part of an intellectual quest to understand how the world works. The Greek intellectual contribution stands apart from all earlier cultures and their relegation of gods to the celestial realm with little or no earthly influence marks an analogous religious break with previous cultures.

Greek science blended philosophical ideas about perfection with observations of the physical world. The Greeks believed that God was the unchangeable Prime Mover who was disconnected from the imperfect earthly world. Historians of science have speculated that the Greek philosophical ideas that served so well to develop astronomical models may have hindered experimentation in the imperfect earthly world. For the Greeks, a designed experiment in which some variables were eliminated or held constant, would so strongly influence the system under study as to make any information gained from the contrived experiment irrelevant for understanding reality. Greek science was pure observation, not experimentation. Astronomy was suited to Grecian study, benefiting from the intellect of many great thinkers who established habits of thought that were to dictate the patterns of inquiry for many centuries.

Hellenistic philosophy, which came to function as a de facto religion, and science were inseparable companions, covering all areas of thought: from music, ethics, and politics, to explanations of the natural world. Astronomical theories represent only a small fraction of the important scientific ideas contributed by the Greeks but following the historical development of astronomy provides a fruitful summary of the interaction between science and religion because theories to understand planetary motions have been central to understanding nature and the origin of the universe.

Perfection was central to Greek thought. Pythagoras (ca 530 BC) asserted that a spherical earth rests at the center of a spherical universe because of the aesthetic attraction of the sphere's perfect symmetry. Pythagoras' assertion that planets and their motion follow spherical ideals is an article of faith, grounded in a level of religious conviction, which can be viewed as an early attempt at stating a scientific worldview. He believed that the heavens were perfect and therefore assumed that spheres and circles were perfect because of the apparent shape and circular motion of the heavenly bodies.

The circular planetary motion proposed by Pythagoras is inconsistent with the erratic movements of several planets across earth's sky. Eudoxus of Cnidus (ca 401–355 BC) proposed one of the earliest systems to reconcile the known movements of planets with their distance and position from the earth. In Eudoxus' model, the sun, moon, and planets are embedded within a series of concentric, transparent, crystalline spheres that rotate around the earth's center. The system accounts for the movements of the planets except for Mars and Venus. These two planets appear to stop their eastward movement across the sky, move in a retrograde westward motion, and then resume an eastward journey. The apparent retrogression is caused by the earth's faster movement so that these planets appear to slow down, stop, and then move in the opposite direction before resuming their eastward journey. Eudoxus cleverly addressed the problem by assigning a second, much smaller and opposite circular motion to each planet which mapped out a non-circular pathway somewhat like that experienced by a rider on the "Octopus" fairground ride. A series of twenty-seven spheres were required to accommodate the various planetary movements while still honoring the ideal of circular motion. The complicated system provided a mathematical model to describe planetary motion but failed to account for the changes in the planets' brightness that results from changes in the distance between the planets and earth.

Aristotle (384–322 BC), who adapted Eudoxus' astronomical system into a physical explanation of true planetary motion, was arguably the most influential Greek intellectual. He believed that the earth was a sphere because an eclipse with the earth casts a circular shadow on the face of the moon and traveling north or south brings into sight stars not previously visible. The celestial bodies were all assumed to be spherical and move in perfectly circular orbits because spheres and circles were idealized. Aristotle's method for developing theories relied on observing, collecting, and collating data from which patterns could be discerned, and extrapolating to universal principles. Aristotle's influence far surpassed any earlier achievements because of his reasoning based on interpretation of data, his extensive writings across the fields of science and philosophy, and his skill as a communicator.

Aristotle assumed that the earth occupied the center of the universe. Surrounding the earth were the moon, Mercury, Venus, the sun, Mars,

Jupiter, Saturn, and finally the stars, all embedded in transparent, crystalline spheres. The crystalline spheres were interconnected through small unrolling spheres that allowed motion from the last sphere to drive subsequent celestial movement throughout the entire universe. For Aristotle, God is the eternal "Prime Mover" who drives the spheres to keep the universe in motion.

Aristotle assumed that, unlike the earth, the stars and planets were composed of a superior element that did not decay or change. He distinguished between a celestial region of perfection lying outside the sphere bounded by the moon, and the terrestrial region extending to earth where change and decay occurred. In this way, astronomical events such as the passage of comets were accommodated in the outer terrestrial region and did not affect the perfect, unchanging heavenly region.

Numerous Greek scientists contributed ideas that refined this basic astronomical structure. Aristarchus of Samos (ca 310–230 BC) measured the angles and positions of planets to determine the distances and sizes of the sun and moon and later Apollonius (born ca 262 BC) added an additional circular planetary movement to account for the changes in a planet's brightness and retrograde motions. Remarkably, Aristarchus proposed the now recognized heliocentric arrangement with the earth and other planets revolving around the sun. Aristarchus was criticized for his heliocentric model, the theory was never accepted, and his only surviving book does not even mention the theory. Movement of the earth clearly violates common sense as anyone that has watched the *sun* rise or set can tell. Aristotle's theory was too widely accepted to be overturned without compelling proof.

Ptolemy (~100–178 AD) was a Greco-Egyptian writer of Alexandria, a Roman province of Egypt, who wrote in Greek and revised the earlier Greek theory of planetary motion. He was undoubtedly one the most important astronomers of antiquity, who refined the earth-centered model to a new level of accuracy. Central to Ptolemy's thinking was his understanding of mathematics as the only sure path to unshakable knowledge. Arithmetic and geometry provided indisputable methods that were more valuable than philosophical ideas. Ptolemy's goal in refining Aristotle's planetary system was to determine a planet's position at any time in the past or future. Ptolemy imagined that a planet's motion around the earth followed a circular pathway that was modified by Eudoxus' epicycles. Each planet's motion was constant relative to a point offset from the center of the earth. The offset changes the speed of a planet during the circular orbit which accounted for the sun's longer duration in the sky during the summer and the retrograde motions of planets like Mars. Despite the model's complexity, the system was remarkably accurate at predicting planetary motion and was used for the next fourteen centuries. Ptolemy's ideas were published in thirteen books known as the *Great System of Astronomy*, later known under the Arabic name, *Almagest*. Ptolemy achieved his aim of creating a regular, mathematical system capable of describing complex, natural phenomena.

SCIENTIFIC DEVELOPMENT IN ROMAN, ISLAMIC, AND INDIAN CULTURES

The Roman Empire is remembered for three crowning achievements: geographical expansion, a comprehensive legal code, and an extensive network of roads and aqueducts unrivaled until modern times. The Roman Empire's world dominance required an extraordinary level of organizational skill. Rome's growth attests to superlative legal and political organizations, whereas the legendary building achievements demonstrate an advanced engineering prowess. Despite inheriting the rich legacy of Greek science and providing centuries of relative peace and prosperity, the Romans made few contributions to the development of science. Rather than using their financial and technical expertise to support science, the Romans became collectors of knowledge, generating handbooks to record important works. Handbooks popularized and disseminated knowledge much like Wikipedia articles do now. Complementing the handbooks were commentaries whose purpose was to explain the meaning of a text. Interpretive remarks were often added and comparisons provided with other authors and commentators.

The relationship of religion to the handbook and commentary traditions was minimal. The cultic emperor worship from 14–337 AD had significant political ramifications and the influence on scientific and technological advances was largely due to political decisions. The subsequent edict which named Christianity as the state religion of Rome in 380 AD likewise had little impact on the quest for new knowledge.

Following the Roman Empire, science developed very slowly during the Middle Ages. Over the same time period, many key Greek ideas were rescued from oblivion in Europe through translation into Arabic by Islamic scholars. Arabic intellectuals subsequently nurtured scientific study for several centuries, making key contributions to mathematics, astronomy, and chemistry. Islamic science benefited from an initial accumulation of knowledge from surrounding cultures—Greek, Babylonian, Hindu, and Christian—using that knowledge first to create encyclopedic collections and then to make scientific advances.

The sweeping religious changes associated with the Islamic conquest of the Middle East and North Africa in the sixth and seventh centuries AD facilitated the assimilation of Greek science into the developing Islamic science. The Qur'an's focus on the omnipotence and oneness of God encouraged the study of nature as evidence for Allah's providence. Despite internal conflicts within the new religion, Islam as a whole displayed a surprising acceptance of ideas from other cultures and faiths which allowed the incorporation of many Greek scientific ideas. Works translated into Arabic include many of the medical writings of Galen, and Ptolemy's astronomy which became best known under the Arabic title *Almagest*.

Islam fostered scientific study as a means to support Islamic doctrine, emphasizing God's transcendence and unity and the affirmation of a creation distinct and independent from God. The Islamic conception of God's omnipotence led to an atomic theory that elevates divine control over all aspects of nature. Arab atomism envisages specific qualities for each atom with God's continual activity sustaining each individual event. For example, God sustains the flow of ink as a pen is guided across the paper to create writing and maintains the final image on the paper afterwards. God could, if he wanted, change the ink color or make the ink disappear altogether if he wished—but because of God's perfect will he faithfully repeats the same causality. The entire universe persists and is maintained only through God's continuous intervention.

The Islamic conception of God facilitated scientific development by encouraging studies of nature that showed God's omnipotence. The greatest Islamic scientist of the ninth century, Al-Khwarizmi, produced astronomical and trigonometric tables as well as a treatise on algebra that provided methods to solve quadratic equations. Islam stifled other areas of science because natural mechanisms were dismissed as explanations in favor of God's continuous creation of events. The most significant contributions of Islamic science were in astronomy and mathematics, areas where observation and abstract thinking were not seen to involve God's intervention.

Although Islamic civilization began to wane around the twelfth century, Islamic culture aided the collection of many foundational works and facilitated several significant scientific advances that were critical for the subsequent development of European science. Much of this knowledge was introduced into Europe through the emigration of educated people following the crusades.

Indian scientists made seminal contributions, primarily to mathematics, over twelve centuries beginning around 400 AD. Indian intellectuals were the originators of the decimal system, introducing zero as a placeholder, and making contributions to arithmetic, algebra, and geometry that were central to the rise of mathematics. Despite Indian scientists making several key conceptual advances, science never flourished as in other cultures. The view of many monistic Indian religions that God is present within all living beings creates a barrier to experimentation because probing living organisms is irreverent. An abstract investigation like mathematics can develop without any obvious interference with nature, but physical investigations of nature are limited by this religious belief.

The brief history of science sketched in the previous sections focuses on astronomy, an area in which religion played a key role. Modern science, distinguished by the use of mathematics and logic to describe the true structure of nature, did not develop in Greek, Arabic, or Indian cultures, leading historians to identify cultural features that might have fostered the development of science in western Europe. Modern science's early beginnings are routinely traced to Galileo's astronomical writings in Italy and the work of Newton and Bacon in Britain, both strongly Christian countries. A cadre of historians have argued that science specifically developed within a western, Christian culture around the time of the Reformation because Christianity held a unique set of beliefs about the structure of the world.

CHRISTIAN BELIEFS THAT FACILITATED SCIENCE

Christianity's gradual adoption throughout much of the Roman Empire enabled the fledgling religion to evaluate, adapt, and adopt ideas from varied cultural and religious traditions. Incorporating the best ideas into a collective framework was particularly helpful in launching a new era of scientific discovery around the time of Galileo. The Christian culture of

Italy at the time of the Renaissance provided a unique environment for science, which is not to diminish significant scientific contributions from other religious cultures, but rather to acknowledge the fact that the beginning of science began in a predominantly Christian Europe. By the late 1500s AD, there were several intellectual ideas embedded within Christianity that had been established over the course of several centuries.

Saint Augustine was an early proponent of accepting both reason and revelation as valid paths to truth. He believed that Greek philosophy and science provided an aid to understanding scripture whereas other truths required mystical revelation. The existence of God and an understanding of God's nature are not accessible by reason but require faith.

Some Greek church fathers, such as Clement of Alexandria, believed that Greek philosophy provided an important preparation for Christianity. Clement argued that philosophy is the handmaid of theology, an image that was adopted by many subsequent Christian thinkers. Other church figures believed that Greek philosophy, such as the Hellenistic writings on wisdom, was nothing more than an extrapolation of ideas contained within the Old Testament. Refining the philosophical ideas of secular thinkers and incorporating the concepts into Christian theology was natural. By the fifteenth century AD, five core Christian principles had been established that facilitated the rise of science.

1. *Creation Out of Nothing*: The Christian conception of divine creation *ex nihilo* contains two implicit assertions: that the universe has a definite beginning and that all of creation, earth and the heavens, have the same origin. Each assertion diverges with Hellenistic thought that saw the universe as eternal with a perfect celestial realm made from an immutable material and an imperfect earth made of imperfect matter.

2. *Creation is Orderly*: Many early cultures viewed nature as wild and difficult, if not impossible, to tame. Nature was often personified as having an independent will with no guarantee that underlying order existed or would be possible to understand. Christians believed that nature had a divine order because God had purposefully created the world. Christians reasoned that if God had created and ordered the world then a study of creation would reveal that order and purpose.

3. *Stewardship*: The creation mandate in Genesis[1] established that care of the garden would provide for physical needs. Christians believed

---

1. Gen 2:15–18

that, like the study of the Bible, scientific study of nature would be rewarded by a beneficial understanding of God. The study of nature was expected to reveal order and purpose that could allow nature to be used in new ways for society's benefit.

4. *Christian emphasis on reason*: Christianity relied heavily on the use of reason as seen in Jesus' teaching, the apostle Paul's intellectual engagement with the Greeks,[2] and the writings of many of the early church fathers. During the time of the Reformation, arguments continually returned to the authority of the Bible. Underlying the accepted truthfulness of the Bible was the assumption that trained theologians, and later even the public, could search for truth and gain knowledge through reading. The individual search for truth was equally applicable to science, and was natural for the many early scientists who were Christians.

5. *Christian tradition of study*: Christianity inherited a tradition of study from the Jews that emphasized scrutinizing the details of scripture. Dissecting Christ's life into parts inspired a cultural acceptance of reducing complex issues into their constituent parts. Science relies on dissecting problems into smaller, more easily understandable, parts a valuable technique known as reduction. The concept of reduction is historically unique to Christianity and Judaism and arguably facilitated the development of science in a Christian culture.

The remainder of this chapter examines how religious views, particularly Christianity, facilitated scientific advances by examining the lives of several scientists whose work was instrumental in shaping the development of modern science. The focus is on astronomy because astronomical studies have been important from the earliest civilizations to the present, but advances in biology, mathematics, and other sciences progressed under similar influences. The cultures for these discoveries were largely Christian, which is not to suggest that Christianity is required for science, but rather to understand how science and religion were perceived by the individual.

The scientific advances made by these individuals were shaped by a complex web of religious, cultural, and personal ideas that guided discoveries in unique ways. The survey of individual scientists progresses from the Renaissance through the Enlightenment and into postmodernity. For some scientists, religion was a powerful motivation, as they believed that

---

2. Acts 17.

God provided circumstances that allowed intellectual breakthroughs. Other scientists found that their work led them to relegate God to an observer or discard religious belief altogether.

NICOLAS COPERNICUS (1473–1543)

Nicolas Copernicus was born into an influential family in 1473. His father was a relatively affluent magistrate in Torun, Poland, who died when Nicolas was ten. Consequently, Copernicus and his three siblings were raised by an uncle. Copernicus' uncle was a canon for the northernmost Catholic diocese in Poland who later became bishop over one of the country's four Catholic precincts. With the help of this benefactor, Copernicus was enrolled in the University of Cracow for an education in the standard broad arts curriculum.

Cracow University was an important center for mathematics and astronomy. The university fostered philosophical questioning, which had an important influence on Copernicus' intellectual development. Copernicus returned home after three years of study to assume the position of canon, which involved doing administrative work for the church diocese, collecting taxes, holding court, dispensing justice, and organizing protection from invaders.

Copernicus' appointment was disputed, leading him to move to the University of Bologna to study canon law for the equivalent of graduate study. His love of astronomy probably helped him become a personal friend of Domenico Maria Novara da Ferrara, a professor in astronomy. Novara (1454–1504) was a Neoplatonist who criticized Ptolemy's complex planetary theory because of his philosophical belief that natural systems were inherently simple. Copernicus aided Novara in documenting star positions and discussed means by which the Ptolemaic system might be corrected. At some point, Copernicus came to the conclusion that Ptolemy's planetary theory was wrong simply because such a complex and cumbersome system was not consistent with the divinely wrought majesty evident in creation.

Copernicus was appointed as a canon but obtained two years leave to study medicine at the universities of Padua and Ferrara. Probably the leave was granted because there were few trained physicians and a trained doctor would be valuable to the church. Copernicus' study of medicine allowed him to continue his planetary observations under the clear Italian skies because medicine was, at that time, understood to be related to astrology. When Copernicus left Italy at age thirty he had an excellent education spanning the classics, law, theology, mathematics, metaphysics, languages, and astronomy.

In 1514, Copernicus completed his *Brief Treatise* criticizing the traditional geocentric system of Aristotle and Ptolemy. The work begins with an attack on Ptolemy's equant theory, which introduces an offset in planetary motion to explain apparent changes in speed of the planets, and proceeds to make a series of assumptions of which the most important is that the sun is at the center of the universe. Copernicus then argues that the apparent retrograde movements of the planets arise from the earth's motion. The book describes details of the heliocentric system, but does not include mathematical demonstrations to support the assertions, which Copernicus planned to explore in a future work.

One year after Copernicus wrote the *Brief Treatise*, Ptolemy's *Almagest* was reprinted. Copernicus realized that a defense of a heliocentric system would require a comprehensive book in order to gain widespread acceptance. While Copernicus continued his astronomical work, he was fulfilling his duties as a canon, initially accompanying his uncle as a travelling companion, private secretary, counselor, and diplomatic representative. Although the intersection of Copernicus' personal faith and his devotion to astronomy may not be immediately apparent, Copernicus' drive to study astronomy came from his understanding that the world was built by the "Best and Most Orderly Workman of all."[3] As a devout priest, Copernicus saw astronomy as a lens through which God's character is revealed in the harmonious structure of the universe. For Copernicus, astronomy was a form of worship.

Despite the ravages of war and the disruptions caused by the Reformation, Copernicus continued to refine his definitive contribution to astronomy, *On the Revolutions*. The preface identified problems in the Julian calendar that could not be overcome until the planetary motions were measured precisely. *On the Revolutions* provided tables devised from a solar (sun-centered) system that allowed an extensive calendar reform in 1582. Copernicus promoted the heliocentric model not only because this gave more accurate results but because he believed this to be the true picture of reality.

---

3. Hannam, *The Genesis of Science*, 277.

Copernicus was well aware of likely responses to a book promoting a sun-centered solar system by a church cleric writing in the time of the Reformation. He was acutely aware of the church's assertion that the earth was the center of the created order because mankind was God's crowning glory. Copernicus addressed these concerns by including a dedication to Pope Paul III that appealed to the Pope's love of learning and especially mathematics. Copernicus reviewed the opinions of several earlier Greek astronomers as additional support for a heliocentric solar system and stressed the harmony of the planets in a system with the sun at the universe's center. *On the Revolutions* was written in Latin as a work for mathematicians which necessarily limited the book's audience and consequently the potential for negative repercussion.

Copernicus' new planetary system preserved uniform circular motion and correlated the distance of the planets from the sun with the period of a planet's revolution around the sun. The order became Mercury, Venus, Earth, Mars, Jupiter, and Saturn at increasing distances from a central sun. The positioning explained Mars' and Venus' proximity to the sun, while the other planets traverse far wider distances across earth's sky. The new system was mathematically elegant *and* physically possible. Copernicus believed that the observations and mathematics described in *The Revolutions* accurately expressed the physical construction of the universe in a far more comprehensive and harmonious manner that represented the true nature of reality.

Copernicus' brilliance lay less in astronomy than in mathematics and in his persistence in refining his heliocentric model to predict observational data. Although the mathematicians who understood Copernicus's book *On the Revolutions* were grateful for the new tables, which made their calculations easier, most of them did not accept Copernicus's theory that the earth moves as a physical reality. A rotating earth in orbit defied what they believed of the physical laws of the time—that a simple body could not have more than one motion—and defied common sense. A moving earth would be expected to leave the birds and air behind and cause a projectile thrown upwards to fall west of the initial starting point.

Copernicus lived in a remote area of Poland without access to a printing press. He therefore entrusted publication of *On the Revolutions* to his protégé Georg Rheticus, a professor of mathematics in Wittenberg. Rheticus' subsequent appointment to the University of Leipzig led to the printing being entrusted to Andreas Osiander, a well-known Lutheran minister and astronomer. Osiander inserted an anonymous preface, "To the Reader," stating that the theories contained in the book were not necessarily true but were to be judged rather by their usefulness at predicting the position of the planets. Copernicus was convinced the ideas were true but Osiander may have thought the preface necessary to protect the book from criticism. Most assumed that "To the Reader" was written by Copernicus which helped the book avoid censure from the church.[4]

Publication of *On the Revolutions* represented something of a time-bomb. Relatively few copies went to Italy and Catholic countries or Britain. Over time and with a second printing the work gained in distribution, promoting a sun-centered solar system to a wider audience.

Copernicus' astronomical work was intimately linked to his Christian convictions. His belief in a sun-centered solar system came from his belief that a perfect God would create a simple universe. This religious conviction was the motivation for pursuing astronomy for two decades in preparation for *On the Revolutions*. Copernicus was aware of the problem a sun-centered universe would have for religious interpretation of several biblical texts but believed that a true understanding of physical reality could be harmonized with the Bible. Copernicus' scientific ideas were directed by his Christian beliefs but little is known of how he thought his science might be used to revise the church's theological position.

JOHANNES KEPLER (1571–1630)

Kepler was born into an unsettled family that lived in a small Lutheran community in southwestern Germany. Religious alliances were a constant source of tension in the area. Catholics and Protestants often lived in close community and some families, like Kepler's, consisted of Protestant and Catholic believers together under one roof. Kepler's fractious father was a mercenary who disappeared from the family records when Kepler was sixteen. His mother was an innkeeper—a strange and unpleasant women with a restless, inquisitive mind Kepler appears to have inherited. Kepler's mother is credited with instilling an early interest in astronomy by taking him to view Halley's comet at age six. The following year he saw a lunar eclipse. Despite a discordant upbringing, Kepler developed a strong faith that influenced his work, perseverance, honesty, and lifelong quest for harmony.

---

4. Hummel, *The Galileo Connection*, 49.

Kepler was a frail boy with constant illnesses which may have helped to spare him from a life of manual labor. Instead Kepler exercised his mind by learning biblical passages by heart which likely honed his intellectual skills. He was an exceptional student who passed a series of highly competitive exams to win scholarships and receive an excellent education despite his modest background.

When he was eighteen, Kepler entered the University of Tübingen, Germany, a famous center of Protestant theology, for a broad education in the faculty of arts. At Tübingen, Kepler was profoundly influenced by his mathematics and astronomy teacher Michael Maestlin who taught Copernican theory to an inner circle of students. Kepler's interest was piqued and he began to seek a thorough understanding of the heliocentric model. At the same time, he earnestly continued his theological studies which led him to question certain Lutheran doctrines, especially the doctrine relating to divine presence in the Holy Communion.

Kepler's genius was recognized at Tübingen and, upon completing his degree, he was appointed as a professor of mathematics in Graz. Although he would have preferred to enter clerical ministry, Kepler later saw the appointment as providential because astronomy provided an unique opportunity to praise and glorify God. In addition to his faculty position, Kepler was appointed district mathematician and further supplemented his income by writing astrological calendars. He believed that the heavenly bodies did influence earthly events but cautioned against using astrology alone. Kepler's astrological predictions, based on his common-sense evaluation of political, social, and economic factors, were remarkably prescient and highly valued.

Initially Kepler's work at Graz went well. Kepler searched for a physical reason that explained why the sun was at the center of the universe, using mathematics as the tool to explain the physical events. Kepler was among the first to establish the scientific axiom that the world is intelligible and that data from observing nature can be analyzed mathematically. Kepler searched for quantitative relationships because he viewed mathematics as the key to unlock the secrets of the universe. Kepler was a mathematical genius, but his belief that Copernicus' sun-centered world truly represented the planetary reality came from his religious conviction that this is the type of beauty that God would create. "I have attested it as true in my deepest soul, and . . . I contemplate its beauty with incredible and ravishing delight. . . . I believe Divine Providence intervened so that by chance I found what I could never obtain by my own efforts. I believe this all the more because I have constantly prayed to God that I might succeed if what Copernicus had said was true."[5]

Kepler taught several topics at Graz but specialized in mathematics. In response to student questions, Kepler noticed a geometrical relationship between the distances separating the planets and the Platonic solids, regular three-dimensional figures with identical equilateral faces. The series of five Platonic solids inscribed within spheres touching at the vertices were of the same relative size as the planetary orbits, evidence that geometry was the archetype divinely woven into the fabric of the universe. The arrangement of tetrahedron, cube, octagon, dodecahedron, and icosahedron surrounded by six intervening spheres provided Kepler with an explanation for the presence of only six planets. For Kepler, the theory was proof of the Copernican model and a testament to God's glory.

Although Graz provided a stable academic environment for astronomy, religious tensions were constantly destabilizing the area. In 1598, Archduke Ferdinand ordered all Lutheran theologians and teachers to leave. This personal tragedy became a formative event in Kepler's life because the events brought him to work with the astronomer Tycho Brahe.

Tycho Brahe (1546–1601) was a brilliant astronomer whose expertise lay primarily in his attention to detail. Under the patronage of King Frederick II of Denmark, he built his own observatory, complete with laboratories, printing press, paper mill, and staff located on the island of Hven in Denmark. Brahe designed and built new instruments that were then used for precisely measuring celestial motion. With the greatest observatory in the world, he was in an ideal position to watch and record the presence of a new star that appeared suddenly, brightened and then faded over the next sixteen months. Brahe's observation of this "nova" or exploding star secured his reputation as an astronomer and, at the same time, provided irrefutable evidence of changes in the heavens. Five years later, a brilliant comet appeared over Europe that was shown to be much closer than the nova—about six times further away than the moon. Brahe interpreted the comet as a challenge to Aristotelian astronomy that precluded changes more remote than the sphere of the moon. Furthermore, the comet's path would have had to traverse through Aristotle's transparent, crystalline spheres, further calling the Greek planetary model into question. Brahe argued that the true planetary model was an immobile earth about which the moon rotated. The planets revolved around the

---

5. Kepler, quoted in Gingerich, *The Eye of Heaven*, 298.

sun which then revolved around the earth. The model satisfied the Dane's Christian interpretation of biblical texts, indicating that the sun, and not the earth, moves, but never received significant enthusiasm from other astronomers.

Tycho Brahe recognized the necessity of comprehensive and accurate measurements to establish the true planetary orbits and had amassed the most accurate record of planetary motions. When Kepler arrived to help Brahe's astronomical program, he was assigned the difficult problem of determining Mars' orbit. Tycho's observations provided the accuracy essential for discovering the true path of Mars but Brahe was overly protective of his observations and limited Kepler's access to the data. After ten months together, Brahe unexpectedly died and Kepler was chosen as the king's leading astronomer.

Access to Brahe's accurate measurements, a newfound freedom to explore diverse interpretations, and a religious belief in the universe's divine harmony kept Kepler from overlooking a relatively small difference between theory and data which other astronomers might have ignored. The discrepancy in the orbit of Mars led Kepler to challenge the principle of *uniform* circular motion in the heavens held by astronomers from the time of Aristotle. Kepler considered his assignment to the problem of the orbit of Mars to be providential because the planet has the greatest orbital eccentricity and was the planet for which Brahe had the most accurate data. Kepler wrote that

> God let me be bound with Tycho through an unalterable fate and did not let me be separated from him by the most oppressive hardships . . . .[6]

Solving the correct path of Mars' orbit was a five-year struggle for Kepler. During his struggle with the motion of Mars, he was encouraged in his belief that geometric hunches reflected divine knowledge that would be rewarded through perseverance. Determining the shape of Mars' orbit required laborious calculations to compare planetary observations with predicted positions. Kepler would settle for nothing less than a theory whose predictions were as accurate as the planetary measurements available. Brahe's planetary positions were accurate to two minutes in an arc contrasting previous longitudinal determinations with acceptable error limits of ten minutes. Kepler's refusal to settle for an accuracy of anything

---

6. Kepler, quoted in Ferguson, *Tycho and Kepler*, 271.

less than two minutes led him, after many trials, to discard modifications to a circular pathway in favor of an ovoid.

The intellectual jump to non-circular motion represented a seismic mental shift from prior astronomers. The break with all previous celestial motion is particularly significant because circular motion was discarded only after a series of tiresome calculations that actually improved the accuracy of the Copernican model, but not to the two minute level of accuracy that was measurable with Brahe's instruments. Kepler began another protracted struggle to calculate and compare Mars' actual and predicted positions which led him to suppose that Mars' orbit was ovoid. Calculating Mars' ovoid path was tortuously complex and took most of Kepler's effort for the year of 1604. Kepler simplified the calculations by assuming an elliptical pathway to model the ovoid path and found the ellipse to accurately describe the actual motion of Mars. Kepler's intellectual path to an ellipse was a repeated give and take between data and theory that is now common among practicing scientists but at the time represented an intellectual break with previous astronomers.

Kepler's insistence on an encompassing, accurate theory was driven by scientific, philosophical, and religious ideals. For Kepler, science and religion were collaborators; a perfect God would make a flawless geometrically-based universe. Kepler believed that Brahe's measurements were a divine gift and the two-minute accuracy was therefore the minimum accuracy for any astronomical theory.

For Kepler, the natural universe was amenable to observation and empirical testing. Although this is second nature to current scientists, prior Greek principles of heavenly perfection led astronomers to propose idealistic models based on philosophical presuppositions rather than discovering comprehensive, realist descriptions of nature through continual refining of the explanatory model. Kepler believed that geometrical qualities were intrinsic within the structure of nature and understandable through mathematical analysis: "Where there is matter, there is geometry."[7] God had imparted numerous principles, generalizability, simplicity, and economy, into nature which required that a realistic description must adhere to the same principles.

Kepler generalized Mars' elliptical orbit into his first law of motion: all planets have elliptical motions with the sun at one focus. Compared to prior models, the heliocentric system with elliptical motion was elegant,

---

7. Kepler, quoted in Kozhamthadam, *The Discovery of Kepler's Laws*, 170.

mathematically simple, and an accurate picture of physical reality. Kepler's model removed the epicycles present in the Copernican system, creating a more elegant series of planetary motions that better accounted for the observed data. Kepler's system explained the variation in brightness of the planets and the greater retrograde movements of Mercury and Venus, compared to the other planets, as a logical consequence of their closer distance to the sun. Kepler also established that the orbits of the planets were in planes passing through the sun at small inclinations relative to the orbit of the earth. The resulting model was about thirty times more accurate in predicting planetary positions than the Copernican model, and provided the first real support, but not proof, for a sun-centered universe.

Studying the movement of Mars on the elliptical orbit led Kepler to his second law of motion, that the orbital velocity of a planet is inversely proportional to the planet's distance from the sun. The discovery is representative of the give-and-take approach Kepler took in his scientific method. Kepler assumed that Mars' speed around, and distance from, the sun were inversely proportional. Describing the relationship mathematically was a challenge that Kepler addressed by calculating the Mars–Sun distance every degree and then determining the distance traversed by Mars through each one degree interval. The laborious calculations led him to apply Archimedes' principle of using triangles to approximate the area between each degree interval. His method was an approximation of the integral method discovered by Newton and Leibniz in the seventeenth century.

Kepler's realist approach to astronomy was accompanied by a similar critical approach to Christianity. His interpretation of the Bible led him to have beliefs most closely aligned to those of Luther but he did not hold to the Lutheran understanding of Christ's presence in and alongside the bread and wine shared during communion. Although Kepler's critical approach oftentimes prevented his acceptance into the Lutheran church, the same attitude likely helped foster Kepler's integrated approach to science and religion.

Religion permeated all of Kepler's life. While he did admit mystery as a component of religion, his ideas were largely guided by rationality. Kepler understood God to have composed the Bible as divine word and nature as divine action. He viewed biblical and scientific studies as being complementary and of equal worth such that his pursuit of astronomy was a spiritual vocation combining the roles of priest and scientist. Kepler's writings were infused with theological reflections because he believed that God's character was reflected in the created world. Kepler often closed his writings with prayer and praise in thanksgiving. He believed that God conceived the very best possible creation that, with geometry as the conduit, captured a divine essence in the universe's structure. Geometry was the tool to understand God's thoughts and unlock the structure of the universe. Being made in God's image meant that each person inherited a mental conception of numbers and quantities that reflected the geometric nature of God.

Kepler saw evidence for the Trinity, the mysterious co-existence of God the Father, Son, and Spirit, throughout the universe's structure. He believed that the central position of the sun was the source of planetary motion in the same way that God the Father is the source of all creation. The sphere, considered the perfect geometric figure, provided another Trinitarian analogy with God residing at the central point, the Son at the surface, and the Holy Spirit residing between. Understanding the universe as having a Trinitarian structure was a guiding principle for Kepler and was apparent in many geometric relationships.

For Kepler, God was the ultimate designer of harmony who had founded the world's structure upon geometry. Pythagoras had recognized the mathematical relationships in nature as early as the sixth century BC, but in the Middle Ages the planetary rotations, driven by an Aristotelian series of linked spheres, were popularly thought to create music: the music of the spheres. Kepler viewed God not only as a mathematician but also as a musician who instilled musical harmony into the world. He sought to relate the musical quality of planetary positions to the harmonics created in plucked strings; only specific ratios of a string plucked once and held at specific positions sound an octave. Kepler worked fervently for many years to discover a collective relationship between the planets and their orbits. Eventually he found that the ratio of the planet's angular velocities at their orbital extremities were in the basic musical intervals: 5/12 for Mercury (octave plus a minor third), 24/25 (minor semitone) for Venus, 15/16 (major semitone) for Earth, 2/3 (perfect fifth) for Mars, 5/6 (minor third) for Jupiter, and 4/5 (major third) for Saturn. Kepler attached deep structural significance to these results because they revealed something of God's majesty in the created world.

Kepler pursued astronomy by searching for harmonious, geometric relationships in the universe's structure. He believed that discovering these relationships revealed God's mind. Kepler pursued apparent relationships for long periods and praised God when he was ultimately able to formulate mathematical relationships. One such search was a twenty-five-year quest to discover a relationship between the position and speed of the planets that eventually led him to find that the ratio of a planet's period squared to the cube of the planet–sun distance was the same for all the planets.

Kepler's approach to astronomy began with a philosophical construct that was then checked against physical data and worked and reworked until a comprehensive, coherent theory emerged. The method, so common to modern scientists, was to develop an idea, an hypothesis, from which deductions could be made and tested by observation. This scientific approach was one legacy left for others to emulate in pursuing scientific discoveries. Throughout his work, Kepler saw himself as a scientist-priest who, through a greater understanding of the true structure of the universe, glorified God through his work: "O God, I am thinking thy thoughts after thee."[8]

GALILEO GALILEI (1564–1642)

Galileo was born in 1564, the first son of an accomplished musician, Vincenzio Galilei. Vincenzio recognized Galileo as being particularly talented, taught him to play the lute, and likely instilled an appreciation for musical experimentation. Despite Vincenzio's musical ability and his contributions to music theory, his financial patronage was modest and sometimes sporadic, which led to interruptions in Galileo's education. When Galileo was eleven, he was sent to study in a monastery at Vallombrosa, south of Florence, where he learned Greek, Latin, and logic. Intellectual life suited Galileo and he offered himself as a novice to join the order, but his father intervened and he returned to Florence.

At the age of seventeen Galileo returned to Pisa, his birthplace, and began to study medicine. His first scientific discovery was made around this time and illustrates his ability to capture fundamentals through observation and to harness the results for practical use. During Mass, he noticed that a chandelier's oscillations took the same time regardless of the distance moved. Working with friends he contrived a *pulsilogium* by which an adjustable pendulum could be set to swing in time with a patient's pulse allowing doctors to know if a pulse was fast or slow.

---

8. Kepler, quoted in Hummel, *The Galileo Connection*, 57.

Galileo developed an affinity for mathematics and considered Archimedes, one of the few Greek masters to combine deduction with practical results, his intellectual inspiration. Vincenzio's financial difficulties forced Galileo to leave the university before finishing his degree and he returned to Florence where he gave private lessons and applied mathematics to various physical problems. During this time, Galileo travelled and developed a cadre of contacts that ultimately helped him to return to join the faculty at Pisa.

As a professor at Pisa, Galileo performed his famous experiment dropping wooden and metal balls from the church's leaning belfry. Although the results lacked accuracy, they directly conflicted with Aristotle's assertion that speed was proportional to weight. Galileo developed his own theory and over his life worked with inclines and planes as a way to control and better understand motion. He overcame the difficulty of accuracy in measuring the time for these motions by devising a controlled water flow, a type of water clock, and weighing the water displaced.

The strategy of intuition, experiment, and demonstration became a hallmark of Galileo's work. Galileo's approach was part of a movement that ultimately changed science from relying on philosophical opinion to a dynamic between observation, theory, and experiment. The emphasis on human senses, and instruments for extending detection such as the telescope, reflected the belief that human reason can detect properties of real objects that can be analyzed according to logical rules to obtain knowledge. This was the beginning of the end for the influence of Aristotelian philosophy on science.

Galileo's demonstrations were augmented with feisty argumentation that earned him the nickname "the wrangler." After three years at Pisa he had offended many colleagues who decided not to renew his contract. Fortunately Galileo received a position at the prestigious university at Padua. The appointment allowed him to support his mother and siblings who arrived to live with him following Vincenzio's death.

Galileo arrived in Padua fifty years after the death of Copernicus and the same year, 1592, in which Friar Giordano Bruno was arrested on suspicion of heresy. Bruno was an itinerant philosopher, thinker, and priest whose belief in a Copernican universe, magic, and pantheism, ultimately led to his death seven years later as an unrepentant heretic. Despite the risks of crossing the Roman church, Galileo's love of controversy and skill in motion and mathematics inevitably attracted him to a heliocentric view of the universe.

The appearance of a new star in October 1604 heralded great change. For a public used to war, famine, and the plague, the astrological appearance seemed to predict ill times to come. For clerics and scholars, a new star was contrary to Aristotle's assertion that the heavens were perfect and unchangeable. Five years later, Galileo learned of a new instrument, the "spyglass," that ultimately allowed him to see with his own eyes that the heavens were far from perfect.

Galileo was in the right place at the right time. Located close to Venice, a premier glass manufacturing center, he was able to obtain several lenses and by trial and error improve the magnification. He attempted to understand the theory behind the diffraction of light to improve the telescope, though the increased magnification came from crafting better lenses. Always an opportunist, Galileo gave one telescope to the Venetian senate, which saw the telescope's value in providing an early warning of potential maritime invasion and rewarded Galileo by doubling his salary.

The importance of the telescope, and a difficulty in securing individual patent rights, saw many versions being made and sold to city officials and scholars alike. In the race to exploit this new instrument, Galileo turned his telescope to the night sky. Galileo saw that the moon was not the perfect sphere Aristotle claimed but had mountains and valleys much like earth. At the limit of his telescope's magnification he discovered that Jupiter had four moons! Quickly Galileo drafted a short book, "Starry Messenger," describing his telescopic discoveries and dedicating the discovery of the four "Medicean stars" to Grand Duke Cosimo II.[9] In return, the Grand Duke appointed Galileo the court mathematician and philosopher, a title Galileo requested to help him change the way people thought about nature.

Not everyone accepted Galileo's astronomical discoveries. Some claimed to look through his telescope and see nothing, while others argued that the images came from flaws in the lenses or were optical illusions. For Galileo, though, the telescope not only provided new data but showed that the universe could be comprehended in a new language of mathematics. Galileo's brilliant intellect, coupled with his fiery rhetoric, proved to be a powerful combination to unleash his revolutionary ideas.

Astronomical observations were slowly causing the Aristotelian worldview to crumble. Using his telescope to project an image on a screen, Galileo refuted Aristotle's claim that the sun was perfect by finding spots

---

9. Moss, *Novelties in the Heavens*, 79.

on the sun's surface. The Jesuit astronomer Father Christopher Scheiner observed the movement of sunspots at much the same time and collected these in a short account. Galileo, regarding telescopic observations to be his own private domain, responded to Schreiner's "Three Letters on Solar Spots" by claiming priority for finding the sunspots and using the drawings to demonstrate that the spots changed shape and were blemishes on the sun's surface. In what was to become a familiar pattern, Galileo alienated not only a potential collaborator but created dissension among many Jesuits whose support would have helped his cause.

Galileo's brusque relationship with religious authorities took a decidedly different turn in 1613. During a court luncheon with Galileo's patron, the Grand Duke Cossimo II and his devout mother, Grand Duchess Christina, a discussion arose about an Old Testament passage from the book of Joshua:

> On the day the Lord gave the Amorites over to Israel, Joshua said to the Lord in the presence of Israel: "Sun, stand still over Gibeon, and you, moon, over the Valley of Aijalon." So the sun stood still, and the moon stopped, till the nation avenged itself on its enemies.[10]

God's miraculous stopping of the sun implies that normally the sun moves, something Galileo held to be false. Present for the luncheon was Father Benedetto Castelli, who relayed the conversation to his friend Galileo. Galileo believed that apparent misunderstandings between cosmology and biblical passages could be resolved. He crafted a letter to Castelli sketching out his views on the relationship between scripture and natural forces. He saw both physical processes and scripture as coming from the same divine source. Any apparent discrepancy arose because the Bible was tailored to human understanding. If a natural phenomenon could be shown to be demonstrably opposed to a biblical passage, then theologians would have to reinterpret the text in light of nature's divine revelation.

Castelli valued the letter which he copied and had distributed. A year afterward, several clerics responded by calling attention to Galileo's foray into theology, a dangerous thing to do in the time of the Counter-Reformation. The matter was taken up by Dominicans at the monastery at San Marco who believed that Galileo had transgressed the Council of Trent: "no one relying on his own judgment and distorting the Sacred

---

10. Josh 10:12–13

Scriptures according to his own conception shall dare to interpret them contrary to that sense which Holy Mother Church, to whom it belongs to judge their true sense and meaning . . . ."[11] Galileo was reported to the Vatican.

Galileo saw the danger and sent an original version of the letter, which differed from Castelli's, to the head of the Jesuits in Rome. Eventually the correct version was evaluated by the Inquisition whose response was cautious, but not unfriendly. Galileo took courage from the response and either did not perceive the political undercurrents or chose to ignore them.

Galileo chose to craft a longer more detailed version of his theological defense as a "Letter to the Grand Duchess Christina." Galileo writes that God has given us the book of nature and the book of Scripture and these "two truths can never contradict each other." He goes on to state that discussions of physical problems should rely on sense-experience and necessary demonstrations and not be made solely on scriptural references. Galileo emphasizes that the primary purpose of the Bible is for salvation and the service of God rather than to teach certainties about the way the world is. He then quotes Cardinal Baronius, saying that the "intention of the Holy Ghost is to teach us how one goes to heaven, not how heaven goes."[12]

Galileo believed that Copernicus was right. Heliocentric beliefs were, in the absence of a definitive demonstration, heretical, but Galileo believed he had proof. If the Catholic Church did not endorse Copernicus's system then he believed Italy's science would be seriously undermined and the Lutherans in northern Europe would triumph. Galileo headed for Rome to win over opponents, rekindle the support of friends, and, if possible, meet with the Pope.

Galileo arrived in Rome in December 1615. The timing coincided with the Holy Office's recalling of *On the Revolutions* for correction, an evaluation of the Copernican theory, and an examination of a personal disposition lodged against Galileo. The time was not propitious for a proof of the earth's movement that Galileo believed could be made from the motion of the tides. He was, however, a celebrity supported by the most influential court in Tuscany. A private meeting was arranged with Galileo and Cardinal Bellarmine, a Jesuit intellectual, and Cardinal

---

11. Machamer, *The Cambridge Companion to Galileo*, 273.
12. Galileo, quoted in Moss, *Novelties in the Heavens*, 201.

Segizzi to convey the Holy Office's position. Although Segizzi ordered Galileo not to discuss or defend Copernicus' ideas, Galileo understood Cardinal Bellarmine to say only that direct support in favor of a heliocentric universe was prohibited. Against this complicated backdrop of alliances and enemies, Galileo returned to Florence.

Galileo wanted to leave a written legacy to rival *On the Revolutions.* Writing progressed slowly on the *Discourses on the Ebb and Flow of the Sea,* the title Galileo decided on for his great work, but after several years he set off for Rome in May 1630 to secure the necessary printing authorization. Galileo perceived that he had powerful enemies, though he now had a long-time friend in Pope Urban VIII. The papal censor was aware of the papal relationship but equally aware that the innocuous title was a fervent and thinly veiled argument for the teachings of Copernicus. After advice from mathematicians, provisional authorization was provided pending several changes including a new title *Dialogue concerning the Two Chief World Systems,* suggested by Pope Urban VIII. As with previous visits, Pope Urban VIII endorsed the protection that a strictly hypothetical dialogue provided. Galileo returned to Florence and began his revision. A year elapsed before the final authorizations allowed printing in Florence with an official *imprimatur.*

Galileo's crowning achievement became a best seller. The book has three men meeting in a Venetian palace to discuss whether the Ptolemaic or Copernican world system is correct. The palace owner, Sagredo, is wise and well-informed, Salviati advocates for a Copernican solar system and Simplicio advocates for an Aristotelian universe. The book is witty, convincing, and encourages the reader to side with one of three characters. Galileo uses many examples from daily life as illustrations to support earth having motion. He argues that motion is relative by showing that when a ship travels through the water; the motion is only noticed relative to the water and islands, not to the items on the ship. In the same way, the earth is not seen to move because all the objects move at the same time with the same speed. After 500 pages that build to the obvious conclusion that Copernicus has discovered the true nature of the solar system, Salviati concludes with an obviously insincere claim that the Copernican planetary system "may very easily turn out to be the most foolish hallucination and a majestic paradox."[13]

---

13. Naess, *Galileo Galilei—When the World Stood Still,* 137.

Copies of Galileo's *Dialogue* arrived in Rome at a time of radical change. Political currents caused Pope Urban VIII to embark on a series of policy and personnel changes to shore up his authority and demonstrate his faithfulness. The shuffle removed several of Galileo's long-time supporters who were close to the Pope. Galileo had several enemies among the Jesuits, who were alienated by his reaction to Father Scheiner's discovery, and among the Dominicans whose favorite philosopher, Aristotle, was under constant attack by Galileo's ideas. Evidence suggests that one of these sympathizers alerted Urban VIII to the concluding arguments of the *Dialogue* in which Simplicio uses the Pope's favorite argument for God's omnipotence to allow for a world to be made "in many ways which are unthinkable to our mind."[14]

Further printing of the *Dialogue* stopped. Diplomatic overtures from Galileo's Florentine dignitaries foundered and a commission of experts was assembled to judge the Copernican character of the *Dialogue*. On September 23, 1632 the Inquisition, Pope Urban VIII, and eight cardinals met to evaluate the charges against Galileo. Galileo was close to seventy when he was summoned to Rome. Despite not being well, the wrangler was intellectually as keen as ever and came before the Inquisition prepared. After establishing the background, the questioning moved to Galileo's visit in 1616 and to guidance from the "Holy Congregation of the Index" stating that a heliocentric universe contradicted scripture. Galileo produced a certificate from Cardinal Bellarmine, now deceased, which explained that Galileo understood the decision of the Index and was not reprimanded in any way. The Inquisition had another letter from the archives of the Holy Office which, though not signed, was likely Cardinal Segizzi's summary of the meeting in 1616. The letter said Galileo was "to relinquish the said opinion that the Sun is the center of the world and immovable and that the earth move; nor further to hold, teach, or defend it in any way whatsoever . . . [and which] Galileo acquiesced and promised to obey . . . ."[15] The two conflicting documents caused a problem.

A panel of theologians was assembled to evaluate the *Dialogue*. Within a few days the members unanimously reported that Galileo had both taught and defended the Copernican theory. After a delay of some weeks, which certainly humbled the aging man, Galileo was clothed in a

---

14. Moss, *Novelties in the Heavens*, 295.
15. Naess, *Galileo Galilei—When the World Stood Still*, 150.

penitent's white gown and led to the convent hall of the Church of Santa Maria sopra Minerva. Galileo was made to kneel and then the judgment and sentence were read.

Galileo was "vehemently suspected of heresy . . . [and therefore] We condemn you to the formal prison of this Holy Office during our pleasure, and by way of salutary penance we enjoin that for three years to come you repeat once a week the seven penitential Psalms."[16] Galileo confessed:

> I, Galileo, son of the late Vincenzio Galilei, Florentine, aged seventy years . . . have been pronounced by the Holy Office to be vehemently suspected of heresy, that is to say, of having held and believed that the Sun is the center of the world and immovable and that the Earth is not the center and moves. . . . With sincere heart and unfeigned faith I abjure, curse and Detest the aforesaid errors and heresies.[17]

Galileo emerged alive but deeply disappointed and in many ways a broken man. What was to be his climatic publication, the *Dialogue*, was placed on the prohibited *Index* of books and Galileo was placed under house arrest. The judgment was not only aimed at Galileo but also his beloved theory, which was forbidden throughout Italy. He believed that he was a good Catholic and had not willingly deceived anyone, especially in publishing the *Dialogue*. Galileo was allowed to leave Rome for Sienna because of the generous intervention of Archbishop Piccolomini, whose lenient housing of Galileo facilitated his eventual return to Florence where he lived for the next eight years under house arrest.

Galileo's fate rested on a complex interplay between his argumentativeness, theology, scripture, and church politics. Galileo appears not to have understood the tremendous change that had occurred in his old friend who was now installed as Pope Urban VIII. Nor did he appreciate the animosity he had created among the Jesuits. The Jesuit mathematician Grienberger has been quoted as saying that if Galileo had kept the affection of the Jesuits, he would have lived gloriously and even have been able to write about the motion of the earth.

Galileo's promotion of the heliocentric solar system was a complex blend of self-aggrandizement, scientific promotion, and desire that the church allow good science and true biblical understanding. He believed

---

16. Stokes, *Galileo*, 173.
17. Galileo, quoted in Hummel, *The Galileo Connection*, 118.

that the Bible was written to describe appearances rather than scientific realities and sought reconciliations for passages that seemed to conflict with scientific theories. He saw God as the author of the book of nature and the Bible, and wanted the church to wisely avoid inappropriate pronouncements from one book to the other. Unfortunately, the politics of the Reformation worked against Galileo and he ended his days as a banished and broken man.

ISAAC NEWTON (1642-1727)

Isaac Newton's birth on Christmas Day 1642 was ironic timing given the unorthodox Christian beliefs he harbored for most of his life. He was the first child of a prosperous Lincolnshire woman whose husband died three months prior to his birth. Newton's mother remarried about two years later, but left her son Isaac to grow up on the family farm with his grandmother. Living without siblings in a remote environment fostered an independent will in Newton.

Newton lived in tumultuous times, surviving the civil war, the Bubonic plague, and the fire of London. The young Isaac became an avid reader and developed a skill in fabricating moveable toys. The mechanical skills he developed would later facilitate his optical and alchemy experiments. At age twelve, Newton entered the Old King's School in nearby Grantham and, after excelling in his studies, was admitted to Trinity College, Cambridge in 1661. Newton's studies included mathematics, Latin, Greek, and the study of optics under his mentor Dr. Isaac Barrow. Newton mastered Kepler's book on optics and went on to grind lenses and develop pioneering theories on light. Using prisms, he split white light into a spectrum of colors and was able to recombine the spectrum into white light. In July, 1668, Newton received his MA and the following year he was installed in the most influential professorship in mathematics, the Lucasian chair in Cambridge.

Trinity was Newton's academic home for thirty-five years and the place of all his scientific contributions. He was a master experimentalist who brought a new type of inquiry to the pursuit of science. Newton's method consists of forming an hypothesis, determining the consequences, and testing the hypothesis by observation and experiment. The critical part of Newton's method was a constant give-and-take between the hypothesis and the experimental results. An hypothesis was refined as a prelude to developing a comprehensive theory.

Newton applied his scientific method to studies that led to new theories on light. Several theories of light were being developed at the time Newton began experimenting with lenses and prisms. Newton's academic appointment required a series of weekly lectures that he devoted to optics and the theory of colors, which helped him to develop his revolutionary theories. In addition to work with prisms, Newton invented and built a reflecting telescope using a curved mirror instead of a lens. The reflecting telescope was much smaller than conventional telescopes made with lenses and produced better images. Newton expected his work and theories to be universally accepted on the basis of logical argument and experimental evidence. Several leading scientists attacked his work, largely out of jealousy, and although Newton's ideas were universally adopted over time, the criticism had deep repercussions. Newton became reluctant to publish subsequent discoveries and some manuscripts were printed only at the urging of friends or to establish priority ahead of his earlier antagonistic competitors.

A statutory requirement of Trinity College was ordination to the Anglican clergy within seven years of beginning an MA degree. Not being one to take commitments lightly, Newton began a lifelong study of the Bible sometime in 1672. He devoured the biblical texts, gaining an extensive command of theology and a familiarity with the writings of the early church fathers. Newton wrote more than 1.3 million words in theological writings that he kept from the public and which were not made available until long after his death. For almost a decade Newton focused on theology and worked very little on the physical theories commonly associated with his name.

Newton pursued alchemy for two decades, believing that he had been specially chosen by God for these discoveries. He recorded his alchemical work in code because alchemy was punishable by death. Newton's goal was to discover the philosopher's stone, a legendary alchemical substance that would turn base metals into gold, because he believed that the philosopher's stone would provide a greater understanding of God. He continued his work in alchemy while writing and developing his better known studies in physics.

Newton's theological studies were interrupted by his work on mechanics, orbital dynamics, and gravity. He developed a theory that for every force there is an equal and opposite force and computed the speed and motion of moving bodies. Newton believed in a sun-centered universe and used his law of gravity to show that the planets maintain their uniform orbits through a gravitational attraction that is finely balanced against an opposing force that causes the planets to move away from each other. Using Kepler's elliptical planetary motion, Newton began the challenging task of reconciling the forces and motion. Ultimately, Newton discovered that the attraction between two bodies varies inversely in proportion to the square of the distance between them.

Despite his pioneering use of mathematical derivations that resulted in new discoveries, Newton was reluctant to publish his work. A young Edmund Halley, who was a friend of Newton, asked him what the relation was between the inverse square law of planetary distance and the motion of celestial bodies. The question had previously been posed as a challenge with prize money for a solution but no scientist had been able to form a solution. Newton told Halley that he had already demonstrated the relationship. Under Halley's encouragement, Newton published a short treatise and then began an intensive study of motion and the attraction between two bodies and the more mathematically challenging multi-bodied systems. For the next year and a half, Newton spent all his waking hours describing mathematical relationships between bodies. Finally, he was able to demonstrate that all matter exerts an attraction to all other matter by a force proportional to the product of the masses divided by the square of the distance between the two masses.

Newton's revolutionary ideas were published in *The Mathematical Principles of Natural Theology*, which is better known by the Latin name *Principia*. Newton's *Principia*, recognized among many scientists as the greatest book ever written, describes the laws of motion and the attraction of gravity. His intellectual leap lay in recognizing that gravity exerts an attractive force between celestial bodies separated by millions of miles of space. The *Principia* was written as a three-part mathematical treatise. Book one described the motions of point masses and their attraction if governed by inverse square forces. The second book refutes Descartes' natural philosophy, and the third book uses the propositions of book one to prove that all planets experience attraction governed by the inverse square law. Although the *Principia* described the forces between bodies, Newton was unable to explain why gravity existed.

The *Principia* is unusual in lacking any history of the universe. Newton believed that the universe was divinely produced as a giant clock-like mechanism. Despite such subtle religious overtones, the first edition of the *Principia* included few religious allusions. Over time, Newton came to see his work as evidence for God the creator and he added theological reflections to the second edition: "This most beautiful system of the sun, planets, and comets could only proceed from the counsel and dominion of an intelligent and powerful Being."[18]

Newton's view of God was dangerously heretical, leading him to confide his beliefs to only a few core friends. Newton came strongly to eschew the Christian doctrine of the Trinity: that God is equally the Father, Jesus the Son, and the Holy Spirit. Instead Newton believed that Jesus was a created intermediary between people and God. Newton's anti-Trinitarian beliefs are ironic given his residence in Trinity College, created by Henry VIII for the study of the Trinity. Aware that Trinitarian beliefs were required of all public servants and for ordination in the Anglican Church, Newton was preparing to lay down his fellowship when his former mentor, Isaac Barrow, secured an intervention through which the holder of the Lucasian chair was granted an exception. For the rest of his life, Newton guarded his secret anti-Trinitarian beliefs from all but his closest friends.

Newton was obsessed with prophetic biblical writings because he thought that verifiable prophetic events provided a proof of biblical authenticity. Just as Newton had formulated "Rules for Right Reasoning in Natural Philosophy," he also formulated analogous rules for interpreting prophetic passages.[19] Using these rules led Newton to predict that the end of the world would arrive in 2060. Behind his work was a confidence that God's message in Scripture and nature were wrought in the simplest possible way. Newton's search for divine simplicity is captured in his many equations in which forces are proportional to simple mathematical quantities.

Newton's extensive theological writings were never published during his lifetime. He believed that his theology was restoring true Christian beliefs. Perhaps because of the need to keep his views secret, Newton's publications laud God much less than do the writings of Kepler. Newton believed that God was continuously and intimately involved in the world to sustain the orderly structure and to serve divine ends. The *Principia* shows how Newton thought that God sporadically intervened in the

---

18. Newton, quoted in Westfall, "Isaac Newton," 155.
19. Olson, *Science Deified & Science Defied*, 129.

world as a motive force. A tract from Newton's final years distills the essence of Christianity to two loves: a love of God and a love of neighbor.

Newton's mathematical analysis of creation led him to see creation as a machine. But Newton did not believe that God made the world and then sat back in his heavenly armchair to watch the clockwork universe; he was not a deist. Newton saw God as ordaining "the creating, preserving, and governing of all things according to his good will and pleasure."[20] Newton speculated that divine attraction was the cause of gravity. Newton believed that God prevented the stars from collapsing in space and would correct the aberrant planetary motions that he had observed from his equations. Whenever Newton did not know an ultimate cause he ascribed that to God.

Newton was appalled with the way God's role changed from sustaining the mechanical universe to becoming irrelevant. Leibniz argued that God's perfection required a universe devoid of constant intervention and within a few years Lagrange and Laplace showed that mathematically incorporating infinite series, rather than Newton's approximations to the first few terms, completely accounted for planetary motion. Philosophically, the move was significant because God's intervention was no longer necessary for the universe's continuing existence. Newton's scientific discoveries were transformed into a mechanistic, god-less, view of the universe that would have mortified him.

Newton's final years were significantly more public than the closeted lifestyle he pursued in his early years of scientific genius. Throughout his life Newton doggedly pursued his work, believing himself to be divinely chosen to discover the universe's deep mathematical harmony. After the strain of a decade of intense intellectual activity, in 1694 he entered into a nervous depression lasting two years. At the end of this time, in 1696, he was appointed as warden of the English Mint. Newton implemented a major recall of old coinage and supervised the delivery of tamperproof coins.

In 1703 Newton became president of the British Royal Society, the main scientific establishment in England. Newton was continually reelected president for the remainder of his life, a position in which he became more autocratic as he aged. The revolutionary scientific contributions of his early years were recognized with a knighthood in 1705 which secured his position within English society as a feted, honorable,

---

20. Newton, quoted in Stokes, *Isaac Newton*, 83.

gentleman. Newton died in 1727 at eighty-five years of age, having written what is arguably the most influential scientific treatise ever penned.

Newton's theories were to dominate physics for two centuries and the philosophical ramifications put in motion social currents whose influence continues to permeate current society. The driving force for much of Newton's work was a belief that he was appointed for God's work, work that resulted in an exceptional series of advances in a short time. Despite Newton's attempts to demonstrate divine providence, his work led to a mechanistic understanding of the universe in which God was not required. Greater insight into the structure of the physical universe appeared to diminish the need for religious belief, a progression repeated during the search for a unified understanding of the biological world.

CHARLES DARWIN (1809–1882)

Darwin was born in Shrewsbury, Britain to a wealthy family with a strong intellectual heritage. Darwin's father was a leading medical doctor with a forceful character who dominated the family home and promoted a materialist perspective. His mother held Unitarian services and was a gentler personality to whom Charles was more attached. Growing up in this vibrant, combative intellectual atmosphere, Darwin inherited a religious sensibility that acknowledged God the creator, but not Jesus' divinity.

Darwin was classically trained at an Anglican grammar school in his early years, which he later characterized as the nadir of his education. Although he did not distinguish himself as a youth, Darwin developed a passion for collecting specimens and through his older brother, Erasmus, a passion for science. Together Charles and Erasmus attended chemistry lectures describing new ideas about the elements and then tried to replicate experiments in their own makeshift laboratory.

When Charles was sixteen he moved to study medicine at Edinburgh University. Edinburgh was a prestigious and thriving university at the time, although Darwin complained about the many dry lectures. They were frequently punctuated with stimulating interactions with professors that engaged Darwin's developing mind. He enrolled in a demonstrative chemistry course in which he learned about the new geological theories that explained the influence of heat on the earth's strata, topics he discussed with his brother. Charles and Erasmus roomed together for Charles' first year and read widely in medicine and natural philosophy which, in the early nineteenth century, was the closest equivalent to modern science. The pair became avid collectors of marine invertebrates that were abundant along the nearby shores of the Firth of Forth.

During Darwin's second year, his interests moved from medicine to natural history. Darwin learned about the classification of organisms, fossils, and geology, and became a professor's assistant helping on excursions to the local beaches and geological surveys. Through these outings Darwin learned of Lamark's new theories describing the adaptive changes in plants and animals brought about through environmental adaptations. Darwin's education in Edinburgh was formative in developing specific interests in animal physiology, bioelectricity, and reproduction, and in instilling a holistic view of nature that was to become a hallmark of his career. Darwin's experience at Edinburgh was also formative in changing his intended career from medicine to the priesthood.

In 1828 Darwin enrolled at Cambridge with the intention of becoming an Anglican priest. Darwin enrolled for a BA degree. Candidates for the BA were required to demonstrate competency in one of the four Gospels or the Acts of the Apostles, and to study the works of William Paley, who provided demonstrations of divine agency through qualities of nature.

While at Cambridge, Darwin developed a passion for collecting beetles and would eventually become a founding member of the Entomological Society of London. He joined a Friday evening group of scientifically inclined students and tutors for discussions which, in conjunction with his reading, kept him abreast of emerging ideas and helped develop his search for pervasive theories in nature. Darwin aided his professors in geological surveys around Cambridge which gave him the practical skills that would later be essential for his work on *HMS Beagle*. The Cambridge experience instilled in Darwin a passion for a comprehensive understanding of the natural world that accounted for the interrelationship between biology, geology, and atmospheric phenomena.

When he graduated, the young Darwin still wanted to enter the priesthood, but his motive had changed from a religious ideal to an opportunity to become a pastor-naturalist. During this time, Cambridge professors recommended that Darwin accompany the Captain of *HMS Beagle* on a surveying voyage to South America. Darwin was to share the poop cabin with Captain Robert FitzRoy, a man with strong Christian convictions who argued with Darwin over his religious views. FitzRoy gave Darwin a copy of Lyell's *Principles of Geology* that was to profoundly influence Darwin during what became a five-year circumnavigation of the globe.

Five years travelling across the open sea in a small, coastal surveying ship punctuated by collecting trips, provided Darwin with many hours for reflection. Most of Darwin's contemplation and writing occurred in transit, which is reflected in the volume of his geological writings that are three times those of the zoological works. The trip caused Darwin to contrast many of the ideas he had assimilated from his education and reading with unrivalled biological and geographical diversity experienced in person. Darwin sought naturalistic explanations for the layering of geological formations, favoring gradualist explanations rather than the catastrophic explanations that fit with many of the religiously informed descriptions of the period.

Darwin's investigations on *HMS Beagle* led him to conclude that a common mode of reproduction unified plants and animals. Subsequent changes among species were related to gradual geological uplift that created a changing environment to which species adapted. He speculated that each species existed for a limited period because the "Author of Nature" had imposed a time-bound life force in each species, similar to the limited fecundity of apple trees. Darwin's personal experience of nature was intimately connected with his religious disposition. On crossing the Andes in Chile, he writes of the experience as "like watching a thunderstorm, or hearing in the full orchestra a chorus of the Messiah."[21]

On returning to England, Darwin decided on a career as a gentleman of science rather than as a parson-naturalist. Darwin had become a highly skilled and creative investigator. His collections of animals and plants won plaudits within the Zoological Society of London, where he enjoyed a warm reception in several scientific circles and took advantage of the availability of journals and books to catch up on the five years of scientific advances he had missed while traveling the world.

In 1838, after drawing up a list of advantages and disadvantages to married life, the advantages won out and Darwin married Emma Wedgewood. Part of Darwin's concern was the health of future children given that Emma was his cousin. Emma, who regularly attended church, was more concerned with relational issues and confided her concern to Charles that his questioning mentality so central to his science might lead him to become skeptical in his religious convictions. The Darwins

---

21. Darwin, quoted in Phipps, *Darwin's Religious Odyssey*, 19.

moved out from London to the countryside and Emma patiently ran the house, protecting Darwin from external influences.

Darwin broke with the tradition of previous naturalists, such as Linnaeus, who were mainly interested in discovering new species, comparing forms, and systematizing the ordering of plants and animals. Darwin viewed himself as a philosophical naturalist searching for comprehensive laws of life that would explain all aspects of earth's inhabitants. He sought to integrate the distribution of plants and animals with geology and the causal process of biological change. In the late 1830s Darwin formulated his theory of descent with modification. Only later would Darwin propose sexual selection as the mechanism.

Darwin's discoveries had troubling religious implications. One of the issues arising from evolution is the nature of humanity. Most of the public held the Christian belief that God had made people in his image through divine fiat, whereas evolutionary theory was providing a plausibly sufficient naturalistic explanation for the development of all living systems. Darwin's theory of natural selection explained that humanity not been divinely created in the Garden of Eden, but had arisen from baser ancestors. Evolutionary theory appeared to undermine morality by providing explanations in terms of survival value that seemed to contradict many Christian doctrines or, at least, ethical perspectives. Darwin's circle of friends included many radicals whose exemplary moral life in the absence of religious belief caused Darwin to question his own views of Christianity. Despite his increasing religious skepticism, Darwin followed advice to avoid or conciliate religion in his writings to advance his scientific ideas. In a private letter, Darwin confides that he wished he had not used the word creation in his writing when he meant "appeared by some wholly unknown process."[22]

Darwin's work was published at a time when natural theology, the argument for God's existence based on design in nature, featured prominently in theology and in training for the priesthood. Darwin wrote that interpreting design in nature as evidence for divine handiwork "which formerly seemed so conclusive, fails, now that the law of natural selection has been discovered."[23] The sheer volume of extinction staggered Darwin. In an early version of his theory, Darwin believed the death and brutality of nature to be justified by the creation of the highest good through

---

22. Darwin, quoted in Phipps, *Darwin's Religious Odyssey*, 94.
23. Darwin, quoted in Larsen, *Evolution's Workshop*, 85.

the creation of higher animals and hints at a rationalization for pain and suffering.

Darwin's religious position, like his science, changed markedly over his lifetime. Darwin increasingly questioned his Christian belief, especially after the death of his father that forced him to confront what he called the "damnable doctrine" of eternal damnation for non-believers. In his *Autobiography* he wrote that "I can hardly see how anyone ought to wish Christian doctrine to be true."[24] He was increasingly troubled by the question of death and how a loving God could allow such brutality and death in the world. For Darwin, the irreconcilable difficulty was compounded by the loss of his favorite daughter which left him angry and bitter at God. Ultimately, Darwin became an agnostic, but described his beliefs as fluctuating, and insisted that there were times when he deserved to be called a theist. Darwin claimed he could not believe that such a wonderful universe was the product of random chance and then, following his rigorous intellectual style, asked why he should trust his own convictions if his mind is the product of a chance evolutionary process. Ironically, for a man of such unorthodox beliefs, Darwin was buried in Westminster Abbey as a national hero and worthy of burial in a bastion of the Anglican faith.

ALBERT EINSTEIN (1879–1955)

On a cold spring morning of 1879, a baby was born in Ulm, southern Bavaria, who was to become the most distinguished of all German scientists. Einstein lived in Ulm for only one year after which his family moved to Munich. Hermann Einstein, Albert's father, partnered with his brother in the developing electrical industry and they became innovative local experts with excellent business prospects. Orders for electrical installations and accoutrements increased and for several years the high-tech firm grew rapidly. The firm peaked with just under 200 employees, but after an unsuccessful bid to secure a lighting contract for Munich city center the prospects for growth became bleak.

As a boy, Einstein enjoyed puzzles, making structures from building blocks and cards. Einstein later speculated that his delayed intellectual growth during childhood might have allowed him to consider childlike ideas in early adulthood which led to him thinking with greater

---

24. Darwin, quoted in Radick, *The Cambridge Companion to Darwin*, 205.

intellectual scrutiny about space-time ideas than most scientists of the time.

Einstein's parents were Jewish but did not attend a synagogue and observed few religious practices in their home. They did not practice kosher rules and ate pork in typical Bavarian style. Einstein's parents enrolled him in a Catholic elementary school in Munich, where he participated in Catholic religious studies as part of the curriculum. At the same time he was taught the rudiments of Judaism by a relative at home. As the only Jew among roughly seventy classmates, Einstein was exposed to the anti-Semitic ethos typical of the culture at large. Einstein did reasonably well at school despite his contempt for an elementary education focused largely on drill exercises. He entered the upper high school and despite wide belief that he was a poor student, his grades were consistently superior to those of most students.

Einstein did not have particularly fond memories of the school, though he did hold several of the teachers in high regard. He enjoyed the Jewish instructor, keenly studied parts of the Torah, and decided to give up eating pork as a consequence of his studies. The young Einstein even composed a few hymns and was planning to enter the Jewish community on his thirteenth birthday, but his decision changed because of the many materialist philosophical books that convinced him parts of the Bible were not true. The experience likely helped Einstein become a free-thinker in religion, characteristic of his later approach to physics.

At age twelve, Einstein developed a passion for mathematics and independently began working through a book on Euclidean geometry. He became proficient in higher mathematical concepts extending from analytical geometry to calculus. About the same time, Albert discovered a great love of music. Einstein had had piano and violin lessons since he was six but in his teens he discovered Mozart's sonatas, which changed music from being a drill to enjoyment.

In 1894 Einstein's father and uncle liquidated their company and moved to Italy because electrical contracts were largely awarded to Germanic, and not Jewish, companies. Albert was to remain in Munich to complete his schooling, leaving the comfortable security of a reasonably affluent family life to live with relatives. Einstein was unhappy with the arrangement and obtained a medical certificate to allow him to leave the school in Munich. He secured a statement from his mathematics teacher stating that he had mastered mathematics up to the graduation level and left Munich to join his parents in Milan.

Einstein promised his parents that he would embark on independent study to prepare himself for entry into the Federal Swiss Polytechnic in Zurich. At the beginning of October 1895, Einstein, at sixteen years of age, had secured an exemption for the entrance exam, which was limited to eighteen year olds, and arrived in Zurich to take the exam. Einstein failed the exam. For the following year Einstein worked to fill in the gaps in his classical knowledge through courses at a high school about thirty miles west of Zurich. During this time Einstein forfeited his German citizenship, probably to avoid being drafted into the German army, and also lost any vestiges of his earlier religious identity. Einstein entered "no religious denomination" in a questionnaire completed in 1900, preferring to describe his religious heritage as an affinity for his "Jewish tribe."[25]

After a year of preparation, Einstein was able to enter the Zurich Polytechnic in 1896. He had two outstanding mathematics professors, though he later admitted to not learning from them as he should. The Polytechnic provided a sound foundation in mathematics and physics with a focus on classical studies. Remarkable experimental discoveries in physics were absent from the curriculum. Einstein intensely followed classes that interested him and skipped others, preferring to study from friend's notes to prepare for the exams. He used the time to study "the masters of theoretical physics with a holy zeal at home," often discovering important works through his friends.[26]

One special friend was the classmate Mileva Marić. Einstein's correspondence with Mileva is characterized by a deep and passionate love interspersed with scientific reflections. A close and unconventional relationship developed that led to the birth of "Lieserl" in early 1902 in Novi Sad. Lieserl was likely placed for adoption or died of scarlet fever; her fate is unclear. For Einstein, a child out of wedlock would have prevented any chance of the professorship that he was seeking.

Einstein graduated from the Polytechnic disappointed with science and did not independently pursue scientific problems for a year. He had hoped to gain a faculty position but this was thwarted because of poor relationships with some of the professors. Instead, he spent a year in short teaching positions with no prospect in his chosen profession. Eventually, in June 1902, through connections with a friend Einstein obtained a position as a patent clerk in Bern. Einstein enjoyed the variation in tasks at

---

25. Calaprice and Lipscombe, *Albert Einstein*, 88.
26. Isaacson, *Einstein: His Life and Universe*, 18.

the patent office and, after settling into the position, found he could complete his assigned work in less than the designated forty-eight hours per week. In later life, Einstein viewed the position as better than a university post because the work avoided the academic temptation to publish superficially and forced mental experiments to gauge the likely success of mechanical machines. The position gave him the financial security to support Mileva and in 1903 the two were wed in a civil ceremony.

Einstein's penchant for independent learning provided the ideal preparation for his miracle year. Einstein had been reading and developing his ideas for some time and established a publication record with several papers in reputable journals before his four seminal papers in 1905. The first of the papers dealt with the energetic properties of light and the photoelectric effect, suggesting energy is exchanged only in discrete amounts or quanta. The theory ultimately won Einstein the Nobel Prize in 1921. The second paper proved that particles suspended in a liquid will have an observable disordered movement caused by thermal motion. The third paper described what was later called the special theory of relativity, while the fourth demonstrated the equivalence of matter and energy through the most famous of all scientific equations: $E=mc^2$. During the same year he published a paper determining the true size of atoms that served as the basis for his doctoral degree from Zurich University. Collectively, the four papers transformed physics and launched Albert Einstein as the world's first scientific superstar.

Receiving the Nobel Prize brought plentiful distractions to the famous physicist who was now courted for pronouncements on social, religious, and political causes. Einstein's humor, candor, and relaxed informality provided a likeable face to the public and, at the same time endeared him as an amiable colleague in the many academic societies to which he was admitted. Einstein travelled to America and Japan and throughout Europe.

During one of Einstein's many high profile trips, he gave an improvised address that provides one of the few insights into his method of identifying and solving intellectual problems in physics. Einstein was able to view unusual relationships between physical reality and the underlying mathematical relationships through ingenious thought experiments. What would the world look like if you were riding at the leading edge of a light beam? What would the hands of a clock look like to a receding observer travelling faster than light? In solving such problems, he described himself as working like a man possessed who was unable to free himself, day or night, to think of anything else. What drove Einstein seems to be an innate passion for science, a passion that helped propel him toward success.

Einstein's problem-solving approach was only influenced by his religious views to the extent of developing simple formulations to describe reality. "What really interests me is whether God could have made the world differently; in other words, whether the demand for logical simplicity leaves any freedom at all."[27]

Einstein was far from the stolid physicist that his work portrayed. His marriage to Mileva began happily, but by 1914 the relationship had deteriorated to the extent that Mileva did not accompany Albert from his post in Zurich to his newly acquired prestigious appointment in Berlin. In 1918 Albert promised the money from the as-yet-unawarded Nobel Prize to Mileva to end the marriage, and in 1919 the couple were divorced. Einstein married his cousin Elsa in the same year. The couple remained together until Elsa's death in 1935 despite Einstein's typically blunt confession that he was not well suited to be a faithful husband. Numerous liaisons occurred throughout Einstein's life which he tried earnestly to keep private.

As Einstein's reputation soared he was increasingly recruited for many causes. Einstein was not a particularly adept speaker but agreed to an American tour whose goal was in part to fundraise for a Hebrew University in Jerusalem. Einstein took pleasure in having served the Zionist cause and ensuring the foundation of the university. Einstein was a pacifist, an ideal that was tested in unusual ways as the world descended into a second war. Einstein was persuaded to write to President Franklin D. Roosevelt outlining the possibility of atomic weapons and recommending the US government begin a program on uranium research. He saw the atomic bomb as an extension of the fundamental problem of societal aggression with the creation of atomic weapons as providing rapid closure to a global crisis. Einstein worked, unsuccessfully, after the war to promote global political unity as a means to prevent future war.

In October 1933, he moved to the U.S. to take a position at the Institute for Advanced Study in Princeton where he worked on a unifying theory of fundamental physical forces until his death in 1955. Einstein suffered from cardiovascular problems beginning in his fifties. In the last

---

27. Galison et al., *Einstein for the 21st Century*, 38.

years of his life, his heart problems increased and rather than choose a high risk operation, Einstein allowed time to take a natural course.

Einstein believed that science and religion benefited from mutual association because their collective search for rational knowledge would benefit humanity. He never believed in a personal God, having forgone religious rituals, even for marriage, and chose not to have a Jewish burial. In his correspondence, he said that he was an agnostic but not an atheist. Einstein believed morality should stem from compassion and he venerated a force of nature, "cosmic religion," that he believed was responsible for the fabric of creation.[28] Religion in the sense of a wondrous awe for nature, was central to Einstein's thinking but had minimal influence on his personal behavior. Pursuing science gave him a conviction of the universe's grandeur and he claimed to have a "spirit of a special sort, which is indeed quite different from the religiosity of someone more naïve."[29] For Einstein, the regularity and comprehensible nature of the universe were hallmarks of divine providence.

CONCLUSION

Tracing the lives and contributions of Copernicus, Kepler, Galileo, Newton, Darwin, and Einstein illustrates how religious convictions and cultural ideals have influenced modern western science's development. The interaction between science and religious belief is complex. For much of history, the two disciplines have existed harmoniously. Religious institutions facilitated the development of science, particularly astronomy, which offered the opportunity to better worship God by accurately identifying the precise timing of feast days.

The Galileo affair is an anomalous instance in which the Church's intervention dampened science's development. This caused a separation between science and Christianity, largely because the Church failed to heed Galileo's warning not to make inappropriate pronouncements in science through a naïve reading of biblical passages. Despite the blot made by the Galileo affair, science and religion have benefited from a continued interaction. The First Vatican Council embraced the insights from science with the declaration that truth cannot contradict truth.[30] At

---

28. Isaacson, *Einstein: His Life and Universe*, 551.
29. Ibid., 388.
30. Vatican, "Catechism of the Catholic Church," 159.

the celebration of the fiftieth anniversary of the Pontifical Academy of Sciences, Pope John Paul II stated that "there is no contradiction between science and religion."[31] Among scientist-theologians there is a general appreciation for theological insight to metaphysical questions that arise from scientific discoveries but which are not scientific in character. Often these are moral questions arising from technological developments such as the development of nuclear processes and their use for military purposes.

Science can provide insight into the structure of the physical world that improves the theological understanding of God. The change from a geocentric to a heliocentric solar system caused a reinterpretation of biblical passages that led to a different understanding of mankind's place in the world, both literally and figuratively. Darwin's theory of evolution meaningfully changed the way most theologians understood the place of humanity in creation. Prior to the theory of descent with modification, theologians saw God as being directly involved in guiding creation. Subsequent to the general adoption of evolutionary theory, several theologians emphasized the divine freedom given for creation to explore biological space. The freedom of the biological world to evolve and develop coincides with themes woven throughout many of the books of the Bible.

The increasingly secularized culture of the last century has seen many great scientific contributions from individuals who claim no religious belief. Religious beliefs appear to be neither a prerequisite for nor an impediment to scientific insight. Unlike Kepler, who saw himself as a scientist-priest pursuing a deeper understanding of God, many scientists see no relationship between belief and the pursuit of science or they see religious belief as detrimental to the pursuit of science. For Darwin, a greater understanding of biology led him far from the priesthood he once thought to join. The suffering that he viewed in nature, combined with several family crises, led him to become an agnostic.

By the time of Einstein, the practice of science had become thoroughly separated from religion. Einstein's motivation for pursuing scientific discovery lay not in religious belief but in an exceptional ability to create mental pictures capable of mathematical formulations. His reverence for nature, his cosmic religion, is distinctly different from the beliefs of Copernicus, Kepler, and Galileo. For Einstein the universe's intricate structure creates a sense of awe and wonder devoid of a personal moral

---

31. John Paul II, Pope, "Truth Cannot Contradict Truth."

imperative. Copernicus, Kepler, and Galileo held God in awe because of their beliefs that were reinforced by their scientific discoveries. Einstein's personal life illustrates the disconnect between his cosmic religion and private conduct as is evident from his effort to erase the details of his many intimate liaisons.

Scientists who are religious may or may not interpret their discoveries as being providential, although for those who do, the process of discovery will likely reinforce their religious beliefs. Religious views offer a motivating force and a way to interpret scientific findings that impacts an individual's metaphysical beliefs such as the interpretation of meaning and purpose in nature. This was true for Kepler who saw design in the planets' order and distance from the sun leading him to praise God.

Just as past scientific discoveries have been influenced by scientists' personal beliefs, cultural and religious views continue to influence scientists today. Two comprehensive studies show that about 40 percent of American scientists profess religious belief, although little is known about how these religious beliefs positively or negatively influence the development of science. In the twenty-first century, the highly specialized nature of science requires an emphasis on logic and rationality that leads many scientists to see God as irrelevant. Historical examples, however, show that scientists' lives and discoveries are uniquely influenced through personal circumstances. Each individual has the opportunity to freely interpret all of life's experiences as experimental data in one of life's most personal interpretations: determining whether or not God exists.

DISCUSSION QUESTIONS

1. Interactions between science and Christianity underlie the historical achievements in many areas of science. Many early scientists were Christians who often occupied clerical positions within the church. On balance, is the historical relationship between science and religion positive or negative? Is this relationship currently positive or negative?

2. Why did the early Hebrews contribute virtually nothing of significance to the birth or development of science?

3. If you were Copernicus, what might you have done to avoid creating a controversy with the church? Was his method of dealing with potential controversy appropriate?

4. Cardinal Bellarmine argued with Galileo that if Copernicus was right then many biblical passages in Joshua and the Psalms would have to be reinterpreted, including a passage in the book of Ecclesiastes: "The sun rises and the sun sets, and hurries back to where it rises."[32] Knowing that there was no definitive proof for a heliocentric universe in the 1600s, what would you do if you were Galileo?

5. Galileo always claimed that he was a good Catholic. Would you agree with Galileo's claim considering his personal life, the style of his intellectual engagement, and his many enemies?

6. Galileo famously defended his pursuit of astronomy by saying that "The Holy Bible and the phenomena of nature proceed alike from the Divine Word,"[33] arguing that both the book of nature and the book of scripture should be read together. Is Galileo's argument valid today, and if so how might this apply?

7. Darwin's idea of survival of the fittest has led to an individualistic focus that is captured in the cultural sentiment of allowing people to do what they want as long as the action doesn't harm anyone else. Within what religious or philosophical framework is this consistent?

8. More than 80 percent of college students describe themselves as spiritual. Do you believe this belief extends beyond Einstein's cosmic religion and if so, how?

9. Is there a legitimate role for prayer in pursuing science? Give reasons for what is or is not appropriate.

10. The history of science is fraught with examples in which sincere religious believers found scientific theories to be at odds with religious views. Although divine truth cannot change and remain truth, the conclusions of both science and religion have changed over the centuries. Give an example of a theological position that has changed because of scientific advances. Give an example of a conflict that emerged in which a later scientific advance showed there to be no conflict.

11. Historically, people have pursued science for religious motives. Is this still the case?

---

32. Eccl 1:5
33. Galileo, quoted in Hummel, *The Galileo Connection*, 106.

Further reading for "A Brief History of Science: From Prehistory to Particle Science"

1. Edward Grant, *Science and Religion, 400 BC to A.D. 1550*, Baltimore: Johns Hopkins University Press, 2006. An extremely readable and yet detailed progression of science and religion from the time of Aristotle to Copernicus.

2. Stanley Jaki, *The Road of Science and the Ways to God*. Chicago: University of Chicago Press, 1980. Professor, scientist, and priest, Jaki argues that a rational belief in a creator was essential for the rise of science. The book is dense but aficionados recognize the work as a seminal contribution to the history and philosophy of science.

3. Richard Olson, *Science and Religion, 1450–1900: From Copernicus to Darwin*. Baltimore: Johns Hopkins University Press, 2004. A companion volume covering the later period of science's development —scholarly in tone and focus.

4. Owen Gingerich, *The Book Nobody Read: Chasing the Revolutions of Nicholas Copernicus*. New York: Penguin, 2005. Imagine trying to find all the copies of Copernicus' book and working out who wrote in them and who they belonged to. Gingerich brings this rich history to life by recreating the sixteenth-century advances and intrigues in astronomy.

5. Charles Hummel, *The Galileo Connection*. Downers Grove, IL: IVP, 1986. A widely read and well-respected introduction to "resolving conflicts between science and the Bible" as promised on the front cover.

6. Jean Dietz Moss, *Novelties in the Heavens: Rhetoric and Science in the Copernican Controversy*. Chicago: University of Chicago Press, 1993. Any understanding of Galileo's influence, science, and religious influence has to be set against his fiery personality. After setting the astronomical stage, Moss captures the fracas by analyzing Galileo's rhetoric in making his points where his science failed to persuade.

7. Atle Naess, *Galileo Galilei—When the World Stood Still*. Heidelburg: Springer, 2005. An excellent distillation of Galileo's life, thought, and influence written in a highly accessible style. Naess captures Galileo's personal character and intellect in ways that ring true to

the complexity of the most high profile drama between a scientist and religious authority. The book is so engaging that the history is absorbed effortlessly.

8. Joseph Spradley, *Visions That Shaped the Universe: A History of Scientific Ideas about the Universe.* Dubuque: Brown, 1995. Spradley aims to bring the concepts of science together with a historical perspective through a focus on mathematics and physics. In ten chapters the book explores science from Egypt through the European Middle Ages and on to modern particle physics to understand how the world came to be and what the key scientific discoveries were.

9 Job Kozhamthadam, *The Discovery of Kepler's Laws: The Interaction of Science, Philosophy, and Religion.* Notre Dame, IN: University of Notre Dame Press, 1994. Kozhamthadam's aim is to show how Kepler's thought was directed by a blend of scientific, philosophical, and religious ideals. Job describes each of Kepler's main discoveries, cross-referencing his publications and his correspondence, to build a picture of what Kepler was likely thinking as his insight developed.

10. James Voelkel, *Johannes Kepler and the New Astronomy.* New York: Oxford University Press, 1999. Voelkel provides a summary of Kepler's life and discoveries. The book is short, thorough, and highly readable.

11. Gary B. Ferngren, *Science and Religion: A Historical Introduction.* Baltimore: Johns Hopkins University Press, 2002. Ferngren has collected the premier writings from his *The History of Science and Religion in the Western Tradition: An Encyclopedia* to create an accessible scholarly introduction to science and religion. Each chapter of ten or so pages captures defining moments in the history of science, beginning with Aristotle and ending with contemporary issues.

12. Jonathan Hodge and Gregory Radlick, eds, *The Cambridge Companion to Darwin.* 2nd ed. New York: Cambridge University Press, 2009. A comprehensive treatment of Darwin's life, work, and intellectual legacy, the book brings together a series of essays by leaders from diverse fields to provide an authoritative text that summarizes key elements in the life and thought of Charles Darwin.

13. William Phipps, *Darwin's Religious Odyssey.* Harrisburg, PA: Trinity, 2002. Williams provides a thorough chronological analysis of Charles Darwin's faith journey. Williams uses the material to

analyze Darwin's lifelong personal religious struggles, avoiding caricatures to provide insights into a complex, intelligent man whose views were not always consistent.

14. Karen Fox and Aries Keck, *Einstein A to Z*. Hoboken, NJ: Wiley, 2004. Written as an alphabetical compendium, the book provides an excellent synopsis of topics in Einstein's life.

15. Walter Isaacson, *Einstein: His Life and Universe*. New York: Simon and Shuster, 2007. One of several comprehensive biographies of Einstein that has recently appeared. Isaacson provides particularly good coverage of key events and themes with an excellent chapter on "Einstein's God."

16. Albrecht Folsing, *Albert Einstein: A Biography*. New York: Viking, 2007. Folsing's tome is very well researched and provides an integrated understanding into Einstein's life and thinking. In several cases, the author compares what Einstein said with others' perspective on his life to draw out the idiosyncrasies inherent in each person's existence, but particularly prevalent in Einstein's life.

7. The Real Me:
Mind, Brain, Soul, and Spiritual Experience

THE HUMAN BRAIN IS the most remarkable organ in all the animal kingdom. The brain has an automatic component that moderates bodily functions, such as heart rate and breathing, and a mental framework that allows willed control over the body in response to focused thought. The brain's mental processing is often compared to a machine because there are at least some brain operations that can be induced. Medical experiments, case studies, and legal and illegal drug regimens demonstrate that there is a mechanistic aspect to brain function that is chemically controlled.

Equally real is the brain's exertion of free will, such as raising an arm or stretching. How immaterial thoughts are able to induce physical movement is an enigma with profound ramifications because the conversion of thought to action is the intersection of immaterial and physical domains. Understanding how a non-physical stimulus can cause a felt response has immediate repercussions for religious beliefs. Religion is anchored in an experienced relationship of physical beings with an intangible divine presence. A very real tension exists between discoveries of the brain's function and understanding how the brain gives rise to consciousness, free will, and religious experiences. Are intangible experiences—whether natural, religious, or chemically induced—real? How should mystical and near-death experiences be understood and what does this mean for religious belief in the existence of God?

Advances in neuroscience are beginning to discover not only how the brain functions but even individual perception and the ability to predict people's thoughts. Although neuroscience is unlikely to locate a

"God spot" in the brain, experiments are probing whether there is an innate tendency to believe in God and a natural receptivity to developing a relationship with a divine presence. Beginning to answer these knotty and profound questions requires analyzing and comparing different models of the human brain.

Understanding how the brain works leads to a natural comparison with computers. A parallel exists between brains and computers but whether the correct relationship is analogy or model is strongly contested. While some neurological processes mimic those of a computer, people generally have a sense of individual awareness that transcends physical processes alone; patients are aware of involuntary reflexes such as those caused by a doctor striking a hammer on a knee. Understanding whether this perception is biologically determined requires parsing the fine details of how the brain might function using a variety of models.

THE BRAIN AS A COMPUTER

Much has been made of the similarity between the electrical processing in the human brain and the mathematical processing of computers. The "Chinese room" thought experiment, however, captures one important difference between brains and computers. The thought experiment imagines a person performing a role analogous to that of a computer. Imagine a person in a room who doesn't know any Chinese and who receives written messages in Chinese symbols, messages that appear simply as incomprehensible squiggles. However, the person also has an instruction book in which each Chinese symbol is referenced to the equivalent English words for the context. When messages in Chinese come in, the person follows the instructions in the book, and writes down the correct translations, which are handed to a messenger, who takes them away. In this scenario, the person functions like a computer following a program: taking the input data, following a set of rules for handling the data, and then giving the relevant output. The point is that the person can do this *without understanding a word of Chinese.* The individual in the Chinese room may appear to an observer to understand Chinese because the translations of Chinese text are correct, but this is an illusion—and here lies the essential difference between computers and cognitive agents. Machines do not understand the meaning of the operations they perform; people can.

A fierce argument rages among artificial intelligence aficionados as to whether machines are capable of evolving to gain a level of understanding independent from that of the original programming. Most people using automated telephone messaging systems would probably say "not in my lifetime!" Deep Blue, the first computer to defeat a reigning chess champion, better illustrates the strides made in artificial intelligence. Playing against a computer capable of evaluating 200 million positions per second, defeated champion Gary Kasparov claimed that the computer's intelligence and creativity was due to human intervention. IBM, the computer manufacturer behind the challenge, denied any human interference.

Kasparov's seminal confrontation with Deep Blue in 1996–97 directly addresses the Turing test for machine intelligence: is a person able to determine whether they are interacting with a machine or a person? Operating within a very large and yet finite set of moves, Deep Blue achieved the same level of play indistinguishable from that of a person. Whether Deep Blue's performance is indicative of human-like cognition by a machine or is more akin to symbolic "Chinese room" processing is hotly contested.

Most people believe that human thinking transcends computer-like decision-making because of their experience with computer systems and reflection on how their own mind works. The distinction between logically-derived understanding and creative discovery has been fruitfully explored in mathematics because the discipline has advanced through a combination of logic and profound creative insight. The field of mathematics generally involves abstract thinking at a level greater than creating mental representations of physical entities for analysis. People are able to identify patterns and understand generalities in terms of fundamental formulas that reflect more than the entities themselves. For example, the Fibonacci series: 1, 1, 2, 3, 5, 8 ... which is obtained by adding the preceding two numbers, can be represented as the formula $F_n = F_{n-1} + F_{n-2}$. Understanding the formula is fundamentally different from a computer calculating a specific number in the sequence. Can a computer derive a similar type of formula without human intervention? Can a computer think? Most people would say that a computer does not know how the mathematical answer is related to the solution output. A computer can calculate large numbers of elements in the Fibonacci series and curve fit data to an equation, but so far still requires human intervention to relate the numbers to practical applications such as growth rates.

A strong argument against computer-based artificial intelligence stems from philosophical work in mathematics. Pioneering work by the mathematician Kurt Gödel proved that mathematics cannot be completely formalized. His incompleteness theory demonstrates that there are mathematical truths that can be demonstrated but not proven within the formal system of mathematics. Basic principles or axioms are required to build consistent mathematical theories and yet, ironically, mathematics cannot be used to prove that these basic principles are true. Conceptually, Gödel's theorem implies that a complete theory of mind may not be possible.

Math is a self-contained system that computers use for complex mathematical processing. Computers are created to "think" in ones and zeros and have not yet been able to make the kind of intuitive leaps in intellectual thinking that undergird the theoretical system upon which these ones and zeros work. The advances in computer technology have created ingenious technological devices that use complex data analysis to provide very human-like output; voice recognition software, for example, provides answers to verbal questions. Such advances obscure the distinction between mathematical processing and human thinking. Proponents of artificial intelligence see the development of human-like computer devices as a prelude to thinking machines that will ultimately allow computers to reach a mental level that will rival the human brain. Others see computers as unable to function without the distinctly human ability to ascend above the processing of numbers to imagine and create new systems. The point of contention is whether a computer's learning ability can supersede that originally programed in by human operators.

Comparing a computer and a human brain reveals similarities in some mental processes. However, a gulf exists in the individual between autonomous bodily processes and the lived experience of top-down, willed, personal choice. Probing the nature of this gulf requires distinguishing between the brain as a central processing unit and the mind as the ultimate controller which, unlike computers, resides in an organic, living, entity. Ingenious neurological experiments are providing insight into the biochemical processes that influence the brain. Understanding the human mind has proven fiendishly difficult because analyzing brains at increasingly sophisticated levels of detail reveals more about the brain's biochemistry but less about the individual. Greater scientific precision is gained by descending from cerebral lobes to individual neurons, but there seems to be an inverse relationship between neuroscientific knowledge and understanding personhood. Insight into physical processes comes at the expense of losing the personal identity inherent in the thought processes. Understanding how the mind operates requires a more holistic approach, scientific and otherwise, to discern the individual experience of mind.

INDIVIDUAL EXPERIENCE OF MIND

During a person's maturation, an intricate network of synapse connections are formed through repeated responses to specific situations. Over time these pathways are activated as natural responses to stimuli. Some of these pathways are automatic while others are perceived as choices in the mind, the part of a person responsible for feeling, thought, perception, will, and reason. Why there should be an experience of choice in the mind is puzzling if the process of cognition is ultimately based on electrical activity caused by chemical processes in response to environmental stimuli or random neuronal activity. Some cells do respond to their environment in predictable ways with no detectable agency; caffeine, for instance, increases heart rate. Somewhere a level of indeterminacy enters between the predictable, determined processes of single cells and the unique responses of an entire ensemble of cells collected in the human brain. The indeterminacy results in people who are complex, unpredictable, and at least appear capable of making choices that vary from individual to individual even under the same set of environmental conditions. This variability among people shows that individual choice is not driven exclusively by environmental cause and effect but is willed by each person's mind.

Almost everyone believes they have a mind: an internal eye capable of making decisions, self-evaluation, and self-reflection. Cognitive scientists observe behaviors that are interpreted in terms of a person's mind—an invisible, intangible, and immaterial object. Amazingly, proving the existence of an individual's mind is impossible despite an essentially universal belief in the existence of the mind across all cultures. From an evolutionary perspective, the ability to infer the presence of a potentially harmful animal or individual, a predator lurking in the undergrowth, might aid in developing beliefs that others have minds with their own intentions, desires, and beliefs. Past explorers observing cultures with no prior western contact found the same basic assumptions about individual's minds, and there is no reason to think that any as-yet undiscovered people groups will have different assumptions.

The very existence of a mind rests on *individual* perception. People are unlikely to have ever seen their brain but generally perceive and believe themselves to have a mind housed *"within"* the brain. The inability to capture images of the mind does not necessarily mean that minds do not exist, but rather that the nature of mind is poorly understood and the tools currently available to probe thoughts are rather blunt instruments.

Most people, on reflection, perceive themselves as more than just complex biological units responding to stimuli. Individuals describe their meaningful experiences not primarily as physical stimuli or bodily responses, but as involving a higher level of cognition, reflection, and decision-making that constitute their unique identity. These higher-level cognitive responses, however, are still deeply tied to the same physical center that controls reflexive bodily responses. Chemically induced brain changes and the differences in individuals before and after a stroke demonstrate a deep connection between mind and body.

The link between mind and brain is elusive in part because there is not a simple correlation between cognition and physical parameters. The anatomical differences in the brains of animals capable of increasingly complex behavior are subtle. For example, brain size does not directly correlate with intellect; intellectual ability correlates with increased brain folds that greatly increases the number of neurons able to be placed in a small space. While mental processing and memory storage occur in the same basic areas of the brain across a broad spectrum of animals the higher cognition among these animals correlates most closely with increased brain connections.

Human brains are remarkably powerful despite the fact that many otherwise likeable people fail to make use of them. People's creative ability seems to outstrip that required by animals trying to live on the savannah. The human brain is much more powerful than the brains of other species, which creates a puzzle if the brain evolved by adding neurons and connections because more examples of species whose brains are capable of complex thought processes would be expected. A highly intelligent human carries an evolutionary advantage for survival, but a brain capable of comprehending quantum physics, crafting complex symphonies, and envisaging mathematical complexities such as irrational numbers appears to have an over-engineered cognitive ability with no obvious evolutionary advantage. A complete evolutionary explanation for the origin of the human mind needs to explain the properties of the mind, why the process leads to thoughts that are true and cohere with reality, the apparent over-developed mental capacity, and how neuronal development and mental programming work together to create the very human perception of mind.

Personhood, the state of being a person with human characteristics and feelings, is the gulf that separates computers from thinking, autonomous individuals. Personhood rests on personal experience and relationships that arise from open personal engagement as described by Kepler's effort to know God in chapter 6. Physical objects, inanimate matter, plants, and animals, can be known by their physical characteristics, but to know a person intimately requires an understanding of the inner thoughts, feelings, and motives that drive physical actions. The understanding of physical characteristics and personhood involve different ways of knowing.

Philosophy provides a complementary way to understand the human mind and integrate personal and spiritual experiences that are not easily amenable to scientific analysis. Philosophy uses thought experiments to develop theories of mind that complement the physical experiments of science that are contingent on studying other brains for information.

Emergence theory is one of the most influential philosophical approaches used to explain the human mind. Emergence assumes that an as-yet-unknown evolutionary mechanism allows a synergistic relationship among the brain's components resulting in mental processing that exceeds the sum of the parts. Triggered by this synergy, the brain becomes capable of sustaining a mind.

EMERGENCE: FROM CHEMISTRY TO COGNITION

The human brain is arguably the most amazing structure in the known universe. With 100 billion cells in a typical brain, there as many cells as there are stars in the Milky Way. Each electrically excitable cell is intricately connected through synapses to as many as 100,000 others to form vast neural networks. These neural networks control biological functions and serve as the platform upon which cognition emerges.

The source of individual identity is intimately linked to an array of electrical signals within the neural networks. In the human body, these electrical signals flow between neurons that are triggered by chemically induced charge imbalances. Metabolism within the neurons allows sodium, potassium, and calcium ions to be pumped across the cell's membrane to create an imbalance of charge which, if large enough, generates a pulse of electricity. Discovering the source of the electrical signals has led scientists to reduce the human brain to the most fundamental units in biology, chemistry, and physics. At the most fundamental level there is no ultimate generator of mentally controlled signals, no brain equivalent of a central computer processing unit. Some people therefore assume that perception of choice is an illusion because a person's actions are ultimately determined by biochemical processes—the random firing of the brain's neurons. The conclusion from this materialist perspective, such as that of the co-discoverer of DNA, Francis Crick, is that "your joys and sorrows, your memories and your ambitions, your sense of personal identity and free will, are in fact no more than the behavior of a vast assembly of nerve cells and their associated molecules."[1]

Thoroughgoing reductionists explain the human mind as the direct result of biological processes: neural activity generates feelings that are perceived as mind through a complex set of interactions that defy simple prediction. Emergence explains the mind arising from a transition in which predictable, deterministic biological processes allow a higher level of cognitive thought where the individual experiences the ability to make free choices. Emergence posits that symbiotic processes allow the brain to be greater than the sum of the constituent components. According to this view, the brain's structural complexity causes something like a phase change which creates a new level of mental cognition resulting in consciousness, freedom, moral responsibility, and even spiritual awareness. The phase change is not a physical change that can detected by MRI, for instance; rather as the neural circuitry becomes increasingly complex, the ability for cognitive function dramatically increases. The goal of emergence is to provide an explanation for the personal experience of mind through a transition from a physical brain to conscious mental states.

Most explanations of emergence assume that high-level cognition is not determined by events at lower levels, but does depend on them. Instead, high-level cognition in some sense controls the lower events such as the firing of neurons to implement mental choice. An individual uses mental programs within the brain, employing cellular mechanisms,

---

1. Crick, *Astonishing Hypothesis*, 3.

to cause a willed outcome such as raising an arm or drinking coffee. The collective neural network leads to the emergence of an individual's mind and the creation of a cognitive agent capable of making choices. Individual neurons are not conscious; a person is.

Consciousness, or mind, is considered an emergent property of the brain that might have evolved in a similar way to that by which babies develop into conscious individuals. Consciousness arises as children develop and interact with their social and physical environments. At no one moment can a transition be identified as the point at which the individual gains a mind or becomes conscious; children are described as naïve or young, but not as having 1/3 or 2/3 of a brain. The basic neuronal architecture appears to be present at birth, which provides the template upon which life experience is imposed; as neurons interact with the world they begin to alter their synaptic architecture, strengthening some connections and weakening others. If the neuronal architecture or the stimulation is absent then normal development is prevented and the individuals fail to become fully functional adults with normal conscious behavior.

CONSCIOUSNESS: WHY IS THERE SELF-AWARENESS?

For humans, the greatest single event in the universe's history, after the Big Bang, is the emergence of conscious, individual minds. Consciousness, arguably a key component of being human, is the perception of being aware of others and of personal identity. Although brain images can show activity associated with increased biological function, the images are not capable of directly measuring consciousness. Sleepwalkers exhibit aspects of attention similar to that of people who are awake and yet have a completely different experience of consciousness. Traditionally, a tacit association of consciousness with human identity was recognized in correlating brain death with loss of all brain activity except that in the brain stem. Now there are moves to declare individuals dead when they lose cognition: personhood is not measured by *biological* life but *biographical* life—variously defined as the sum of personal aspirations, decisions, activities, and human interactions.

Conscious biographical life is central to each person's existence. Normal awareness is created in the brain's left and right hemispheres, which broadly inform logic and emotion, respectively. Cooperation between the left and right brain hemispheres is intimately tied to mind.

Surgical cleavage of the connections, sometimes used in treating epileptic seizures, can lead to patients experiencing two separate minds. For patients with disconnected left and right brain hemispheres, an object placed out of sight in the left hand does not appear to match an identical object placed in the right hand.

Surgical procedures, analyses of individuals with brain abnormalities, and brain mapping are unable to correlate one discrete area of the brain with consciousness. Where then does consciousness come from? Some parts of quantum theory assume that consciousness actually lies *outside* an individual. Quantum mechanical equations require an observer outside the system of observation both as a way to judge the description's accuracy and to isolate the system from the influence of the investigator. The observer is the fact-finder, the evaluator of facts. This places the evaluation and mental processing of the observer's mind outside the physical realm of observation. From this perspective, a person's mind lies to some extent *outside* the possibility of mathematical and physical explanation.

Neuroscientists trying to locate the origin of an experience or idea face several challenges in identifying areas in the brain that create desired outcomes. The "binding problem" is the apparent ability of the brain to take information from diverse stimuli and to interpret them as a single experience. Viewing an object causes specific cells in the eye to register color, shape, and movement, and yet the brain registers everything collectively as one unified object. Adults experience themselves as single, unified entities, both at any given time and over time. A conversation with a friend is not just a series of disconnected stimuli—the curve of lips, vocal sounds, and visual shapes—but is collectively interpreted by the brain as an entire input and understood as a scene with an emotional resonance. There is some higher-level processing that binds stimuli together in a way that has relevance for understanding the experience of the mind.

Human consciousness is necessarily subjective because individuals both experience consciousness and are able to reflect on their own consciousness. People have the remarkable ability to imagine tangible and intangible events in their mind and to recreate the same image without even having the object present. A person can close their eyes and picture a light bulb in their mind, but there is no light source in the brain and no internal "me" to see the imagined light. Recent brain scanning techniques show identical brain patterns in individuals viewing an object, such as a light bulb, as when imagining the same object in their mind. Even observing or experiencing a situation and mentally recalling the same situation leads to brain patterns that are essentially indistinguishable. These experiments highlight the difficulty of relating physical images to the mental operation of an immaterial mind.

Perception is sometimes traced to specific brain regions by asking patients what they feel to correlate these feelings with active areas of the brain. Arriving at a unified understanding of what constitutes a person's mind is complicated because each person has a unique experience of only their own mind. The source of the overall sense of consciousness is unknown; consciousness may arise because of an interaction of the component parts or through an as yet undiscovered mechanism without specific symbiosis of the sub-components. Perhaps consciousness, like gravity, is a fundamental component of the world that is not amenable to detection through dissection into smaller components.

The idea that consciousness is the same in all people is inferred but not proven. Personal experience is a thorny issue because each person's perception is different. Favorite colors, food choices, habits, personalities, and attractions toward others vary from person to person with the more meaningful choices having greater the variability. Studying people's understanding of their own consciousness provides insight but does not reveal the underlying mental operation. External monitoring of the brain can detect brain function but fails to detect consciousness. Here lies another dilemma in neuroscientific research: accurate information requires the experimenter and the human subject to be the same.

Among the most personal choices are an individual's religious beliefs. Some research indicates that far from entering the world with a neutral perspective, young children are naturally wired to accept belief in God. The beginning of morality in infants less than two years of age has been linked to bonding with their caregivers. Attachments between parents and the infant, such as crying or chewing bring the parent to the child while crawling brings the child to the caregiver, both of which promote proximity. Over time these natural attachment tendencies are erased as the infant explores further from the parental place knowing that support will be available when needed. Parents are a source of security because they are available when needed. Researchers suggest that these early memories of attachment experiences prime individuals for an innate religiosity or spirituality. An equally reasonable way of understanding infant developmental behavior is that these patterns of belief would be exactly that expected to understand and believe in God.

Pre-school children who were asked to sort objects according to whether they were made by people, God, or unknown, were seven times more likely to identify God as the maker of natural objects than people. Children five through ten, regardless of religious background, favored a creationist account for the origin of animals regardless of their parents' teaching on evolution or divine creation. Children saw living and non-living natural objects as having a purpose, leading many psychologists to accept that children have a bias toward belief in creation of the world by a non-human super being.

Most people reevaluate their religious beliefs during the transition from childhood to adulthood. Belief in God or the absence of a god is decided through a deeply individual set of experiences that underscores the unique, and incompletely understood, nature of the human mind. The personal choice between belief and non-belief may be influenced by crises, the religious views of the prevalent culture, or religious experiences achieved through meditation or through added physical stimulus.

MIND-ALTERING DRUGS

During the religious rituals of some indigenous Amazonian Amerindians, shaman prepare a drink containing DMT, a hallucinogen-inducing alkaloid, to facilitate divine experiences and aid healing. The exact role of DMT, a mind-altering molecule optimistically called the "spirit molecule," remains incompletely understood. DMT, N,N-dimethyltryptamine, is a derivative of the amino acid tryptophan that is rapidly degraded by the body's various defense systems in the gut, bloodstream, and brain. DMT is an unusual psychedelic because the chemical is found in numerous plants and animals as well as the human body.

The use of mind-altering drugs reveals direct physical connections between the body, the brain, and the mind. Being tired, drinking alcohol, and taking narcotics makes concentration harder, increases the risk of mistakes, and can lead to unusual behaviors. The origin of the perceptual imbalance caused by psychedelic drugs is often due to interference with receptor binding to serotonin, a tryptophan-derived neurotransmitter involved in many physiological functions including sleep, mood, and cognition.

Early research with psychedelic drugs was focused toward relieving mental illness by chemically stimulating or repressing areas of the brain responsible for specific actions. High-profile personal use, particularly of LSD, led to stringent laws governing the use of psychedelic drugs, largely because of the dangers of an altered perception of reality: visual and auditory hallucinations, a distorted sense of time and space, and an enhanced sense of meaningfulness. People who take hallucinogens have visions, hear sounds, and feel sensations that cannot be observed by others.

Interpreting these chemically induced experiences is extremely challenging because an individual's interpretation of the experience can only be observed indirectly. Recreational hallucinogens produce highly variable effects, in part because of the variable quantities in different plants but also because of individual physiological differences and the influence of environment. Subjects interpret these events in diverse ways, from a psychedelic trip to a period of spiritual transcendence. Evidence suggests that the psychological effect of these drugs depends on the person's expectations, the setting, and the worldview from which the experience is subsequently interpreted.

Psychiatric medication harnesses the link between mind and brain by chemically stimulating or blocking biochemical processes to induce brain states that make the individual feel "normal." Antidepressants, such as Prozac, operate on well-known biochemical machinery, the central neurotransmitter serotonin and various receptors, which regulate a host of physical and psychological processes from heart rate to mood. Prescription treatments again illustrate the individual nature of brain chemistry; psychiatrists usually prescribe a low level of treatment based on past experience, choosing the precise medication and dosage by closely monitoring a patient's behavior and their own feelings. Psychiatric medical intervention provides valuable treatment for depression, mood alteration, and related psychological conditions that are typically augmented with psychological techniques.

Various psychotherapies such as cognitive behavioral therapy can have a profound influence over an individual achieving a healthy mental state. The sustained successes of Alcoholics Anonymous-type addiction programs demonstrate the power of an immaterial will to exercise control over the body. Behavioral programs such as Alcoholics Anonymous help individuals exert their free will to control thoughts and influence their behavior. Ironically the first steps of all Alcoholics Anonymous-type programs are to admit to not having control over addiction or compulsion and recognizing that a higher power can give strength. The success of twelve-step recovery programs focuses attention on precisely what free will is and why belief in a higher power is an essential component.

FREE WILL

Free will is intimately linked to the mental capacity to understand abstract concepts and judge their truthfulness, and then to make decisions that are not solely determined by physical processes. Traditionally, free will is thought of as being free in neither a random nor a deterministic sense, but captures the personal experience of being completely autonomous moral agents capable of an internal will to make unrestrained choices.

Experiments performed during brain surgery substantiate a will independent from mechanistic control. Areas associated with speech have been mapped out prior to surgery by asking patients to identify a series of pictures by name. Patients were unable to name the pictures when electrodes were applied to the area where the speech cortex was supposed to be. Patients tried unsuccessfully to use related words, accusing the experimenter of stopping them from retrieving their words. In other patients, electrical stimulation was applied to cause a patient's hand to move. Invariably the patient would insist that they did not move their hand but the surgeon did. Some patients even used their other hand to stop the electrically induced hand motion. In no instance did electrical stimulation in the cerebral cortex cause a patient to believe or to decide anything. These brain experiments, performed early in the twentieth century, imply that there *is* an independent controller, a person's free will, that is able to activate the brain's circuitry to cause bodily responses.

A person's independent ability to cause bodily movement implies that the body is not *only* controlled by material causes. The possibility of a non-material influence on free will is consistent with the quantum description of the physical world at the smallest and most fundamental level. The unpredictability of nature is captured in statistical models. A world of "probably" rather than "certainly" provides an opportunity for an external influence, mind.

Physically, the "seat of the will" is most tightly associated with the brain's prefrontal cortex, the region of the brain governing complex movement and goal-oriented processes. Brain damage to the prefrontal cortex causes a loss of concentration and increased difficulty in planning and carrying out complex mental tasks. Individuals with damage to the frontal lobe tend to be emotionally flat, indifferent, and lack volition. The case of railroad worker Phineas Gage illustrates the profound character differences that can occur from damage to the prefrontal cortex.

Phineas Gage was a twenty-five-year-old railway foreman who was described by his New England employers as "most effective and capable." As he prepared an explosion to remove a block obstructing the path of the railroad, someone called to him from behind and he briefly looked away. He turned back and began tamping the explosive with his iron bar, without realizing that his assistant had not poured in sand beforehand. A deafening explosion took place. The bar entered Gage's left cheek, pierced the skull, crossed the front of his brain and emerged through the top of his head. The rod landed more than 20 meters away, covered in blood and brains. Phineas Gage was stunned but amazingly still conscious. He went on to make a remarkable recovery, but his personality changed dramatically. His likes and dislikes, his aspirations, his ethics and morals were all different.

Gage's story highlights a unique feature of the brain: unlike other organs that can be surgically altered, have portions removed, or even be replaced, physical changes to the brain can cause dramatic mental and neurological changes. At the extreme, brain death leads to a loss of life. In Gage's case, significant behavioral changes resulted from damage to the prefrontal cortex, the area of the brain which is largely dedicated to reasoning and particularly to the social and personal dimensions of reasoning. These areas are associated with thought patterns unique to humans: anticipating and planning for the future, responsibility, and exerting free will.

Moral decision making is often broadly discussed in terms of making conscious, rational choices where the personal and situational outcomes are weighed before taking action. Experientially, moral behavior often occurs seamlessly without a conscious reasoning process, implying that at least part of morality stems from habitual responses that constitute a component of an individual's character. Damage to the frontal cortex, where memories are often stored, decouples the conscious and automatic decision-making processes so that behavior anchored in previous experience cannot be recalled. In these cases, individual behavior can become erratic and capricious.

Medical advances are unmasking the intimate connection between free will and brain function. Neuroscientific analyses have focused more on deterministic processes with direct cause and effect outcomes rather than the more challenging area of understanding personal choices. For instance, research has shown that anti-social behaviors can be caused by brain malfunctions and even diet if the correct balance of vitamins, minerals, and protein are not available. In 2000 a schoolteacher was arrested for child molestation after being unable to stop himself from visiting pornographic websites, collecting sexually explicit magazines, and making advances towards his stepdaughter. An MRI revealed an egg-sized tumor whose removal decreased his lewd tendencies and returned his normal sexual behavior. Interestingly, the incident revealed that the man had a tendency toward inappropriate sexual content but was able to restrain his inclinations. A year later the man's immoral behavior surfaced again, which led to further brain imaging that showed a partial return of the tumor. Removal of the tumor again reoriented his sexual behavior to normal. The relationship between the teacher's behavior and the tumor pressing on the prefrontal cortex demonstrates the link between moral behavior and the brain; proper functioning of the brain is required for the exercise of free will.

Functional magnetic resonance imaging is able to detect brain events that allow an individual's subsequent choices to be predicted with remarkable accuracy. For example, patients engaged in economics games exhibit distinctive brain patterns that allowed researchers to predict the person's next move. There is considerable uncertainty in the prediction, though even 100 percent accuracy would not mean that people lack choice. The question is what causes the brain state from which the prediction is made. At some level a free choice must be caused by a physical process and/or an immaterial decision; freedom and causality must be compatible, a philosophical position called "compatibilism."

Individuals can exert free will to resist natural and unnatural inclinations, wants, desires, and cravings not only because of fear of being punished or the desire for a reward, but also because of moral convictions. People make decisions based on their own preferences, rather than merely reacting in a predetermined, programmed fashion. Evidence for freedom of choice is partly supplied by the number of people who make different choices in the same situation. Heroes may defy "normal" choices because of their high moral standards. Martyrs give up self-preservation for a religious cause.

The examples of heroes and martyrs show a connection between free will and moral choices. Most religions and the legal system assume that each person is responsible for her or his actions. The law presupposes a free will that individuals are able to control to overcome deterministic tendencies; people are expected to overcome sexual instinct to conform to cultural norms. Legal punishment is partly intended to change individual choice. Juries have rejected the argument that "my genes made me do it" in favor of personal responsibility. Free will provides an escape from the materialistic chain of cause and effect through a meta-level judgment where good and evil can be weighed. A world with true choice requires people having the option of choosing evil over good, as discussed in chapter 4. Consistently exerting the will to choose good creates neural pathways that make these good actions not just choices but also habits, establishing a moral character that cannot be inherited like biological traits.

Many people's free will choices are guided by a belief that people's lives will be judged according to whether they led a good or bad life. There is widespread belief that if God existed then his main job would be to judge people's lives once they die. Although such beliefs are likely to be strongly influenced by cultural norms, there have been sporadic examples of individuals returning to life and reporting visions of heaven that seem to confirm God's existence. Spectacular advances in medical technology have increased the frequency of resuscitations and the occurrence of near-death experiences. Near-death experiences provide a fascinating set of indications that indicate that life is not limited to the physical world and that the most important aspects of life are actually immaterial. Although near-death experiences are necessarily personal and anecdotal, medical advances are increasing their frequency to the point where their collective force becomes compelling.

NEAR-DEATH EXPERIENCES

Near-death experiences elicit near universal fascination because everyone wants to know what happens after death. People who have recovered from bouts with death describe similar experiences: they feel themselves leaving their body, traveling through a tunnel, and coming into a heavenly world where they are met by beings of great light. Often the person experiences a rapid review of their life, meeting people in a world of light, and being given a choice whether to go back to life on earth or not. On choosing to return, an individual feels sucked back, a feeling often accompanied by great pain and distress. In many cases the experiences are life-changing, leaving people without a fear of death or with a diminished concern for material things, and with a greater concern for spiritual life and personal relationships.

In many near-death experiences, people claim to be aware of the immediate environment during the time of this out-of-body perspective. Patients who experience remote viewing often describe scenes that have later been independently verified: descriptions of an ambulance, car wreckage, an operating theater, clothing and jewelry, and people present at the scene of an emergency. In one case, an unannounced family visit to the hospital was described by a patient who was undergoing resuscitation and would otherwise have had no knowledge of the visit. In another instance, remote viewing was recorded by a blind person.

Several patients who have had near-death experiences, particularly patients who suffered cardiac arrest, have been connected to machines monitoring brain waves which registered no activity. Over a five-year period, cardiologist Dr. Michael Sabom collected more than thirty cases of patients with serious heart disease who claimed remote viewing of a resuscitation procedure performed on them in the operating room. The patients' knowledge of the resuscitation procedure was compared with twenty-five seasoned cardiac patients who had been consecutively admitted to a coronary care unit. The seasoned cardiac patients' descriptions each contained at least one major procedural error whereas the patients experiencing remote viewing gave details that closely corresponded to the actual hospital procedure without the same type of major errors.

Creating a comprehensive understanding of near-death experiences is fiendishly difficult. Collectively, these experiences seem to provide empirical evidence that consciousness and perception can survive outside the brain. If consciousness can exist in an afterlife without a physical body then the core of being human must lie in more than just a physical body. Philosophy and religion have been trying to probe this core of human identity for centuries, often by reference to the human soul. Over time the view of the soul has progressed from an intangible soul inhabiting a body to a modern unified understanding that defies simplistic disconnection into two components.

THE SOUL—THE REAL "ME"

Most people, religious or not, have a pervasive belief that a personal self, a "me," a soul, survives after death. While people realize that death is the end of the body, the mind has a difficult time projecting *not* thinking, and *not* acting onto the dead body. Projecting agency onto living people is a natural part of life. Belief in a soul becomes a very easy progression from projecting agency. Some even suggest that the human mind is wired to believe in the existence of souls.

Historically, each person was seen as having a physical body and an immaterial soul—the dualist position. Saint Augustine's dualist idea of a "rational soul using a body" served as the basis for the early church's position.[2] Thomas Aquinas subsequently argued that the soul exists independently, apart from the body, so that an individual is a soul-body composite. Dualists view the soul as the organizing principle of the body, directing all bodily processes from the smallest functions of cell metabolism to the most ephemeral ideas generated in the mind. From a dualist perspective, the soul must affect the physical processes of the body; the challenge lies in understanding how a seemingly immaterial soul influences a material body. One analogy is to imagine body and spirit intertwined like DNA; both strands are essential for physical life but conceptually the spirit strand could be separated and re-read like a computer download. At death, the soul leaves the body to provide a template for installation into a resurrected, renewed person in much the same way as computer programs are reproduced and reloaded into new machines.

Several ingenious—and highly speculative—solutions have been proposed to explain how the soul persists after a physical death. In most cases, these solutions assume that the soul empowers the individual with essential human characteristics paramount of which is the ability to relate to God. A different view of the soul is that brain functions allow all mental ability, including the ability to have relationships with others and with God, but that a spiritual relationship depends on God's supernatural initiative. Near-death experiences pose a problem for this position because the immaterial soul appears to exist without a living body.

Catholic theology currently teaches that every soul is bestowed by God sometime between the moment of fertilization and the following two weeks. The imprecision in the institution of the soul accommodates theological conundrums such as the origin of two unique souls in the

---

2. Stump and Kretzmann, *The Cambridge Companion to Augustine*, 116.

case of identical twins who separate from a single fertilized egg. The Bible doesn't directly teach that people have souls but rather presupposes the existence of souls. God breathes *life* into Adam. Within the Hebrew Scriptures, the soul is the life and vitality instantiated into a person's being. Biblically, people are pictured as souls whose lives are maintained by the spirit of life breathed into the body at creation by God. At death the spirit leaves. The key factor in understanding human nature from a biblical perspective is not any physical or functional trait, but the "image of God." This is how the Bible distinguishes people from animals; humanity is the only being whom God created in his own image. The second chapter of Genesis describes God forming Adam from the dust of the ground and breathing "into his nostrils the breath of life," to form a "living being." The soul is the "real me," which for Christians means both corrupted by sin and redeemed through God's sacrifice. People experience their soul through interactions with people, nature, and God.

Modern theological opinion is united in affirming that God's image is relational, not physical. The Bible describes men and women as individuals living in relationship to each other and God. The behavior of people made in God's image affects their relationship with each other and with God, even though tendencies in human behavior may have an entirely naturalistic explanation. In other words, the essence of being centers on an individual's relationships more than their physical body.

Regardless of the origin of the soul, the religious value of the soul is dependent upon an evolving relationship with God. The greatest commandment given by Jesus was to "love the Lord with all your heart, mind, strength, and soul."[3] Humans have value not because of something they have: a gene, a personality trait, or the way they look, but because of a unique relationship with a loving God.

The resurrection of Jesus Christ provides a prototype for understanding the nature of a reunited soul and body. Jesus's body arose from the tomb and appeared to the disciples complete with the marks of the crucifixion. Jesus's resurrected body was able to pass through locked walls and doors, unlike the bodies of human beings, but was immediately recognizable as Jesus's body. The resurrected Jesus ate and drank but more importantly, Jesus's character and memory was continuous. Nowhere is this more apparent than in his relationship with the disciples, specifically Peter who had earlier denied any knowledge of Jesus under questioning.

---

3. Luke 10:27; Matt 22:37; Mark 12:30; Deut 6:5

Jesus took pains to reestablish a personal relationship with Peter and ensure that Peter knew that his mission was to have relationship with fellow believers. Traditional Christian theology states that each individual will experience the same type of resurrection as Jesus on the ultimate Day of Judgment.

An attractive explanation of individual resurrection is that each person's unique identity survives in the mind of God because of the relationship established over the course of a person's lifetime. The relational information allows for a divine recreation of the essential core of a resurrected individual in a form recognizable by others. The perpetual replacement of the individual atoms in the human body, demonstrates that personal identity is tied to information rather than a distinct bodily form. Any physical resurrection of the body based on currently known biology would lead to an individual subject to the same types of limitations, imperfections, and decay suggesting a very different resurrected form that will reside with God forever. Jesus's resurrected body is the prototype, bearing recognizable marks of his crucifixion and yet of a different form from his human body.

RELIGIOUS, SPIRITUAL, AND MYSTICAL EXPERIENCES

Traditional views of the soul as the center of a person's spirituality and moral core are being re-examined because of the recent ability to chemically or electrically induce mental states through specific activation of brain regions. Magnetic resonance imaging is being used to collect brain images of people engaging in religious activities to try and understand the physiology associated with religious experience. The images show activity in several different brain regions for complex mental processes that suggest that the experienced soul comes from an interconnected web of various mental states rather than activity in one central "God spot."

People having spiritual experiences have specific patterns of brain activity, which is to be expected because the experiences are perceived and interpreted by the mind. Interestingly, while electromagnetic stimulation of a person's temporal lobe does not induce a spiritual experience, the likelihood that the person will have a spiritual experience does increase during stimulation. Whether the likelihood of a spiritual experience increases because of a heightened receptivity to a spiritual dimension, or whether temporal lobe activity causes people to think there is a spiritual life is under debate.

Surveys show a remarkably high percentage of people who categorize themselves as spiritual. Spirituality concerns a person's inner, subjective life, personal values, beliefs about the purpose of life, and connectedness to the world and other people. Spirituality includes many experiences that are often not reflectively developed in a direct, logical, fashion—intuition, inspiration, the mysterious, and the mystical. Spiritual people are expected to exhibit heightened emotional and moral characteristics such as love, compassion, and equanimity.

Recent studies show that people become more spiritual during their college years. A direct correlation exists between spiritual growth and improved academic performance, psychological well-being, and leadership. Actively viewing life as a spiritual quest heightens a sense of connectedness which promotes ethical behavior, responsibility, empathy, and social justice. A seven-year study of students showed that measures of spirituality increase for students engaged in interdisciplinary studies, study abroad, service learning, charitable giving, interactions with other races, leadership training, and contemplative practices such as prayer and meditation. Spirituality is linked with, but doesn't require, religious belief, although generally religious people are expected to be spiritual.

Accompanying the high self-identification of individuals as spiritual are the incidents of religious or spiritual experiences. Between 20–49 percent of the population report such incidences. Scientific descriptions of religious experiences in the early nineteenth century focused on a correlation with temporal lobe epilepsy: a disturbance of electrical brain function in the temporal lobes located just above the ears. However, people with temporal lobe epilepsy often exhibit other symptoms not observed in people having a religious experience.

Although temporal lobe epilepsy is not currently held to be the source of spiritual experiences, the search for physical correlates raises an intriguing question as to whether the experiences themselves are physically or mentally induced. Electrically stimulating the brain, or suffering an epileptic fit, can cause patients to have experiences similar to those of people having spiritual and mystical experiences. In one sense, these electronic stimuli are like chemical stimulants, alcohol and illicit drugs, that also cause perceptual changes in the brain. Individuals recognize the difference between physically and mentally induced experiences, in part because each person trusts their perception of the world and has come to believe that their mind's perception is true. The question remains: does the brain receive a real spiritual influence from an external being?

Most people develop their spiritual ideas from a collection of unconscious assumptions about the way the world is. Social scientists have devised experiments that suggest that people readily assume that *something* is responsible for visual images that are not clearly perceived. A small brown object moving in the distance is automatically assumed to be an animal. Large footprints in the woods are ascribed to a yeti. Crop circles, the geometrical patterns that can appear overnight in grain crops, cause some people to attribute their formation to God or aliens. Movies and books have used the themes of coincidences and signs to explore faith and belief. When all the lights turn green, people may wonder if God is responsible. Once a person assumes that God is responsible for producing these signs, the mind reinforces the belief from a store of foundational beliefs held in a person's memory.

Some researchers interpret religious experience as arising from changes in brain activity. The specific areas of the brain responsible for spatial orientation are often decreased in individuals having a spiritual experience that correlates with a feeling of being in harmony with the world as a whole. Other researchers argue that while brain activity is expected to change during a religious experience, the key interpretation depends on prior religious experience combined with interpretation of the event. Temporal lobe seizures can cause similar experiences to those of a religious state but without the religious interpretation. Studies with twins suggest a genetic component to people's tendency for a religious experience. Identical twins' transcendent experiences are more similar than those of fraternal twins, as are their religious practices such as church attendance.

Claims have been made that a "God circuit" or a "God module" has been found in the brain. The explanation has been advanced that such a module is an evolutionary adaptation that might help bring order to society. One neuroscientist has gone so far as to develop a "God helmet" which applies weak magnetic fields to the right and left sides of the brain to induce a spiritual experience. Reportedly, over 1,000 people have tried the God helmet with vastly different results. Self-confessed atheist Richard Dawkins experienced nothing except twitches in his leg but others had powerful spiritual experiences and visions. Researchers at Uppsala University in Sweden were unable to reproduce the results and suggest that people donning the God helmet were predisposed to a mystical experience because of environmental cues performed before and during the experiment.

An individual having a spiritual experience is primarily concerned with understanding the meaning of the experience. For most religious mystics the aim is to know God more through an intensely spiritual state of consciousness. Precisely because religious experiences are subjective, interpretation is particularly challenging. The best scientific answers can only define what spiritual experiences are *not*. These experiences are not hallucinations. People who experience hallucinations understand that the perception is not real, whereas individuals having spiritual experiences believe that they have had an encounter with God. Following a religious experience, a person's subsequent behavior is often profoundly affected.

Different studies have tried to identify what happens to the brain during mystical experiences. Neurological monitoring of Tibetan monks and Carmelite nuns entering a peak state of mystical union identified over-activity in some areas and suppressed activity in other areas that help locate the physical correlate of meta-physical experiences. Brain activity in the orientation association area, the area that perceives the limits of the body, is dramatically suppressed during mystical experiences. The correlation fits with individuals perceiving a timeless unity and diminished influence from their surroundings.

In another study, brain scans of Franciscan nuns and Buddhist meditators showed decreased activity in the part of the brain that helps orient spatial position and a simultaneous increase in activity in the frontal lobe for increased concentration, attention, and focus. During prayer or meditation, the nuns feel a sense of being close to God while the Buddhists feel a sense of timelessness. Scientists argue about whether the brain activity causes individuals to believe in God, or whether divine interaction causes the individual to experience God's presence through specific patterns of brain activity. Any experience of God would be expected to be perceived by the brain with a scientifically detectable correlate, which makes this debate unresolvable with the current level of technology.

The experiments identify specific biological responses to religious experiences which should not be confused with the brain having a "God spot." Unlike language, where precise neural systems and structures can be located in the brain, different religious experiences correlate with up- or down-regulation in different parts of the brain depending on the nature of the experience. Understanding how an immaterial God interacts with physical beings at the causal joint is a challenge, for science in understanding the mechanism and for religion in developing effective practices.

Religious beliefs vary tremendously. Some people believe in a God in a detached fashion while others believe that God consistently interacts in their personal lives. For most believers, personal experience forms the key evidence that God not only exists but also engages in an intense and personal relationship. For these people, the surprise is not that spiritual or mystical experiences occur but rather that they occur infrequently. For many, even a one-time encounter with God can cause dramatic changes in beliefs and practices.

CONCLUSION

After the decade of the brain, from 1990 to 2000, and the presidential "Brain Initiative" launched in 2014, the brain remains one of the most complex of all structures in the known universe. The brain is both like and unlike a computer, capable not only of logical processing but also complex thought. Perhaps the biggest gulf between a mechanistic model of the brain, based on computational or biochemical models, and free-thinking individuals is human consciousness. Organic physical processes ultimately lead to individuals with minds that are affected by physical —chemical and electrical—stimuli but which give rise to free will and the ability to override biological instinct. Despite the power of free will, scientists cannot prove the ability of this simple, non-tangible aspect of life that influences behavior.

Scientific imaging is providing a powerful insight into the biochemistry of the human brain and a measure of the perturbations that occur during religious engagement through prayer, meditation, and mystical experiences. What evades detection is an identification of the soul, the real me, and the thought source. The incomplete understanding of the brain, particularly in explaining how the mind arises, should not be a license to propose an unknown God module in the brain. However, there is a direct parallel between tangible matter giving rise to intangible thought and physical bodies being able to sense an intangible divine presence. The inability of scientific instruments to detect mind and soul is unsurprising; what is surprising is that the effort has led to so much understanding—much of which supports the religious views of the historic Christian faith.

Surveys showing a high number of people having spiritual experiences raise the question of whether these encounters point to a world pregnant with opportunities to connect with a divine presence. Despite near-death experiences, which provide a particularly intense personal encounter with obvious spiritual implications, more often the event causes survivors to focus their remaining lives on relationships rather than increased religious involvement. Near-death experiences, mystical experiences, free will, and the operation of the mind collectively point to relationships with God and people as being among the most important aspects of being human.

DISCUSSION QUESTIONS

1. Experiments indicate that physical areas of the brain are responsible for specific functions. Reductionist descriptions of neural firing may not be sufficient to describe the experience of mind or even to identify the position of the mind in the body. Do you accept a difference between mind and brain? Is mind explainable solely in terms of brain components?

2. Natural selection is based on trial and error, refining information to gain a better system. Inherent in this process is the idea of never achieving the ideal system but slowly progressing closer to the best system. Does natural selection provide a sufficient criteria for the formation of a mind capable of having certainty?

3. What is a human being? Does the answer to this question first require an answer to the question of what happens at death?

4. Advances in modern medicine now allow bodily functions to be maintained in people with minimal brain function. A distinction has arisen between bodily life and personal life which stops in people categorized as being brain-dead. Should these individuals be categorized as brain-dead patients or are they biologically sustained corpses? Is there a difference between a person and a human?

5. A long history of neurological patients with brain disorders has identified individuals who are partially or completely incapable of appropriately regulating their own behavior. These individuals may

violate social conventions of civility, ethics, or even governmental laws, which places them at risk in their personal relationships and even as free individuals in society. Is a person without any ability to moderate their behavior "soulless"?

6. Have you ever been mentally troubled from stress, grief, or anguish without being ill? Do you think this is caused by physical or spiritual distress to your brain, mind, or soul?

7. Have you ever experienced God's presence?

8. Do you believe there is life after death? Is your answer based on philosophy, religion, experience, science, or something else?

9. Do descriptions of near-death experiences simply push the limits of consciousness by a few hours or for eternity?

10. Scientific and religious ideas are based on information from a variety of sources ranging from "hard data" to revelation and intuition. Have you ever found faith in an idea necessary to truly understand the fullness of a situation or individual? What strategies for understanding do you find most useful for knowing tangible and intangible ideas?

11. People have speculated about the existence of life in other parts of the galaxy and generally have conceded there is no theological reason why God might not have created life on other planets. Assuming life was found elsewhere in the galaxy, how would you determine if any alien sentient beings had souls?

12. Imagine a century in the future where advances in genetic and robotic engineering allow a genetically engineered dog to talk to a robot. How would you determine if either one has a soul? Do animals with relational capacity have souls?

13. Often as a person ages their mental faculties diminish, particularly for those suffering from Alzheimer's disease, which necessarily impinges on their spiritual and relational capacity. What do you think is the effect of old age on a person's soul and is there any effect for the individual in an afterlife?

Further reading for "The Real Me: Mind, Brain, Soul, and Spiritual Experience"

1. Stephen Barr, *Modern Physics and Ancient Faith*. Indiana: University of Notre Dame Press, 2003. In part V, Stephen Barr, a physicist at the University of Delaware and a Catholic, analyzes "What Is Man?" from physical and metaphysical perspectives.

2. Daniel Kahneman, *Thinking Fast and Slow*. Reprint. New York: Farrar, Straus and Giroux, 2011. Kahneman received the Nobel Prize in 2002 in economics for his groundbreaking work on decision making. *Thinking Fast and Slow* describes the two basic thought processes that control peoples' lives pulling from extensive research in the field of behavioral economics.

3. Alexander W. Astin, et al., *Cultivating the Spirit: How College can Enhance Students' Inner Lives*. San Francisco: Jossey-Bass, 2010. The book summarizes a seven-year study examining what develops and hinders spirituality during the college years.

4. Terence Nichols, *The Sacred Cosmos*. Grand Rapid: Brazos, 2003. Written from a theological perspective, the focus is on God's interaction in the world. Particularly interesting are the chapters on human nature and the soul which rely on observable phenomena to support the widely held belief that there is more to the universe than the visible world.

5. Justin Barrett, *Why Would Anyone Believe in God?* Walnut Creek: Altamira, 2004. Barrett analyzes modern advances in cognitive science to understand where belief comes from. Materialist and evolutionary perspectives are included without judging the veracity of underlying assumptions, but rather focusing on explaining where beliefs in God come from.

6. Deborah B. Haarsma and Loren D. Haarsma. *Origins*. Grand Rapids: Faith Alive Christian Resources, 2007. Provides an excellent introduction from a reformed perspective. Different positions are presented in a non-judgmental manner showing advantages and challenges faced by each position.

7. Rick Strassman, *DMT: The Spirit Molecule: A Doctor's Revolutionary Research into the Biology of Near-Death and Mystical Experiences*. South Paris, ME: Park Street, 2001. In this strongly supportive

description of DMT and psychedelic drugs, Strassman describes his five years of DEA-approved research on the effects of DMT. While disturbing, the personal accounts of near-death experiences, mystical experiences, and alien encounters provide a unique insight into chemically induced changes in the brain. Despite twenty-five of sixty healthy volunteers reporting from minor to serious negative experiences, Strassman remains confident in the potential use of DMT.

8. Christopher Baglow, *Faith, Science, and Reason: Theology on the Cutting Edge.* Chicago: Midwest Theological Forum, 2009. Baglow deftly focuses on the philosophical issues emanating from the intersection of science and religion. He writes from a Catholic perspective and quotes liberally from church figures with an emphasis on Catholics.

9. Steve Stoller, *The Symphony of Creation: Science and Faith in Harmony* Phoenix: ACW, 2002. Using musical metaphors, Stoller answers many of the main questions in science and religion for a general audience. Stoller, a medical doctor, focuses particularly on the personal relevance of these issues.

10. Joel B. Green and Stuart L. Palmer, eds., *In Search of the Soul: Four Views of the Mind-Body Problem.* Downer's Grove, IL: IVP, 2005. Four philosophical positions on the mind-body problem are analyzed by Christian philosophers. Each essay is followed by a response from the other contributors, keeping the arguments focused and highlighting the distinctions in the positions under discussion.

11. James P. Moreland and Scott B. Rae, *Body and Soul: Human Nature and the Crisis in Ethics.* Downer's Grove, IL: IVP, 2000. The authors thoroughly scrutinize the nature of body and soul from a philosophical perspective. In the two parts of this book, the authors first survey different philosophical positions with historical background and then a rebuttal. In the second part, the authors examine a series of ethical issues and develop or examine the logical stances based on different philosophical positions.

12. Malcolm Jeeves, ed., *From Cells to Souls—And Beyond: Changing Portraits of Human Nature.* Grand Rapids: Eerdmans, 2004. An excellent series of essays by a combination of scientists, philosophers, and theologians that write on the relevance of the soul for their field.

The book contains numerous examples of the difficulties that arise in trying to pinpoint what being human means, particularly with reference to patients with Alzheimer's disease.

13. Malcolm Jeeves and Warren Brown, *Neuroscience, Psychology, and Religion: Illusions, Delusions, and Realities about Human Nature*. Conshohocken, PA: Templeton, 2009. Two neuropsychologists provide an overview of the main issues in mind/brain/soul that provides the background for a holistic framework to accommodate emerging discoveries in neuroscience. The book is particularly valuable in summarizing complex advances in neuroscience in an accessible way and identifying the key issues and responses.

14. Kevin Seybold, *Explorations in Neuroscience, Psychology, and Religion*, Farnham, UK: Ashgate, 2007. Seybold provides a particularly helpful introduction to modern neuroscience and psychology for the non-specialist that then forms the basis for examining the religious implications of recent scientific advances.

8. Where Science and Religion Meet: Is there Personal Relevance?

THE RELATIONSHIP BETWEEN SCIENCE and religion has changed significantly over time. The chapter begins by examining models that describe their interaction and then evaluates whether the effort of pursuing an integrated approach to science and religion has any personal value. The chapter closes by briefly identifying recurring themes that have arisen in the chronology from the Big Bang, chapter 1, through to the inner life of the contemporary individual, chapter 7.

THE WARFARE MODEL OF SCIENCE AND RELIGION

Popular contemporary culture often depicts religion and science locked in unalterable and irreconcilable conflict. The historical record shows the opposite: science largely developed because of religious quests to know more about God through the study of nature. The idea that science and religion are inimically separate and antagonistic is rejected by virtually all historians of science. So how did the understanding of science and religion at war arise and become so pervasive?

Most scholars of science and religion view the warfare motif as originating from changes in the public perception of science beginning in the middle of the nineteenth century. The rise of Darwinian theory caused a deep rift in established religious ideas and resulted in a cultural shift that viewed independent enquiry through science as the ultimate path to truth. As a result, natural law replaced providence as an explanation of natural phenomena, society placed an increased emphasis on scientific interpretations rather than religious exegesis, and people began to separate ethics and morality from theology.

Two authors, John William Draper and Andrew Dickson White, were extremely influential in causing this change. John Draper was the son of a Methodist minister who received a medical degree from the University of Pennsylvania and went on to become an influential chemistry professor. Draper argued that natural science was the liberator of religion because he regarded Catholicism as a pretentious, oppressive, and authoritarian form of religion that was preventing people from realizing their full potential. In 1874 he collected his ideas in a *History of the Conflict between Religion and Science*. Draper stressed the promises of a better life through scientific advances which, in the late nineteenth century, resonated strongly with the public.

The book was wildly successful, going through more than fifty printings. The polemical style found a sympathetic audience despite the fact that the assertion of conflict rested largely on an unreliable historical foundation: arguments supported by poor references to previous scholarship, quotations taken out of context, and a dubious synthesis of historical facts.[1]

Two years after publication of the *History of the Conflict between Religion and Science* came a second book with the same theme, *The Warfare of Science* by Andrew Dickson White. He was born in 1832, raised Episcopalian, became an eminent historian, and was subsequently installed as the first president of Cornell University. As president of the first university to be established without a religious affiliation, and with an entirely secular program of study, White wanted to show that religion restrained science's development. He elaborated on the theme of science as the liberator of academic freedom in a two-volume book, published in 1896, called *A History of the Warfare of Science with Theology in Christendom*. White's books had copious footnotes giving the impression of a scholarly approach to the topic but the historical accuracy of many of these footnotes is suspect. For example, a passage asserting that people widely believed in a flat earth is supported by citation to *A History of the Life and Voyages of Christopher Columbus* by Washington Irving.[2] The problem with this citation is that Irving's work is clearly a fictional account. Although White did not personally believe that science and

---

1. Principe, *The Great Courses*.
2. Ibid.

religion were enemies, his publications provided a compelling case for the warfare concept because he was committed to advancing the secular educational and research program at Cornell that he believed was stifled by religious influences. The combination of White and Draper's books established an urban legend with an easily remembered metaphor that has stuck.

Over the following century, science's prestige grew tremendously while the influence of religion waned. Religious believers have, at times, tried to stop the perceived encroachment of science into religious issues, usually with the result of reinforcing the warfare image. For example, attempts by Christians to limit the teaching of evolution and promote religious alternatives have been struck down by the courts on more than one occasion. The Scopes Trial of 1925 arose as a litmus test for the teaching of evolution in Tennessee's public schools with backing from fundamentalists on one side and the American Civil Liberties Union on the other. The teaching of evolution resurfaced in a Pennsylvania court in 2004 which ruled that the Dover Area School cannot "denigrate or disparage the scientific theory of evolution"[3] or promote intelligent design. Intelligent design (see chapter 1, "Complexity and Design") was deemed not to be science.

Some scientists have promoted a separation between science and religion by asserting that science is the only reliable method of determining what is true. These scientists distinguish between the testable, experimental observations of science and the unique nature of religious beliefs based on individual experience and tradition. The issue with this perspective is not that science affords a path to certain kinds of very important truth, but that science provides the *exclusive* path to any and all truth; that outside of science there is no truth. Moreover, statements that only science provides a sure path to truth are philosophical assertions rather than scientific conclusions. Most importantly, this position excludes many of the most influential, personal features of the lived experience.

3. Young and Largent, *Evolution and Creationism*, 288.

THE SEPARATE SPHERES APPROACH TO SCIENCE AND RELIGION

The separate spheres model emerged in response to the conflict model of science and religion. Also known as the non-overlapping magisterial approach, the model is rooted in an absolute separation between each magisterium, or domain of authority. In science, the experimenter strives for detachment whereas in theology a personal encounter with material and spiritual elements is essential for religious belief.

The separate spheres model contrasts the objective basis of science, which questions the physical nature of the world, with the spiritual experience of religion, which answers personal questions about meaning, purpose, and destiny. In each case the authorities are different; for science the authority is logic and experimental observation whereas religion rests on divine authority revealed to key religious individuals and validated in personal experience.

The separate spheres model avoids contention by defining two exclusive domains which necessarily excludes exploring issues at the intersection of science and religion. But engaging the two domains can facilitate insight into perplexing scientific and religious questions. Einstein identified one such enigma when he said that "the only incomprehensible thing about the universe is that it is comprehensible."[4] There is no reason why the world should have the order and rationality on which science relies; the universe could be chaotic or have a rationality that is wholly inaccessible to human thought. Although hominid evolution requires a certain congruence between the functioning of the human mind and the earthly environment that influence the mind's development, Christian theology's assertion that the human mind is infused with the same divine structure as creation provides a metaphysical answer to why the mind can perceive the universe's mathematical sub-structure. The perspective is not evidence for a creator; rather the stance adds insight consistent with a theistic worldview. The greater coherence of this religious perspective and the ability to develop similarly broad philosophical perspectives, are possible only by transcending the simplistic separate spheres approach.

---

4. Polkinghorne, *Theology in the Context of Science*, 72.

THE INTEGRATION OF SCIENCE AND RELIGION

Science, and the technologies that come from science, are inadequate to address the attendant philosophical, moral, and religious issues. Issues such as chemical enhancement of physical and mental performance, fertilization methods with multiple embryos, the development of remotely piloted warfare, and the sustainability of earth's limited resources, have important moral consequences that are beyond the scope of science. The tools to work through these complex issues can be found in philosophy and in religion. Religion has the potential to purify science from presumption and provide answers to the metaphysical questions that arise from science but which are not themselves scientific in character. Religious insight can guide science and technology for the good of mankind, as is evident from by the establishment of hospitals and the development of medicine over the centuries.[5] Such guidance is needed because while science is based on logic and analysis, human nature is not. Religion can raise individuals above their biological instincts by promoting lifestyles and choices that can help moderate scientific advances that might be achievable but not necessarily wise.

Science and religion are susceptible to error because both are human endeavors; fortunately both are corrigible. Science has a history of informing religion about the way the world is that has purged errors from theology and removed superstition. For example, science was central in the long journey from a geocentric to a heliocentric solar system and facilitated changes in the way theologians approached the interpretation of biblical texts. In a similar way, the central Christian tenet that each person is made in God's image was challenged by Darwinian evolution. The subsequent evaluation of the doctrine has resulted in a richer theological understanding of humanity and the value and vagaries inherent in the human condition. Theological reflection on these issues has led to general principles such as the sanctity of human life that is used to guide scientific research on the earliest stages of human life. Difficult points of intersection will continue as science progresses: advances in neuroscience have triggered a reappraisal of traditional views of the soul and challenge previous theological speculation describing how an immaterial divine power sways human will. Science and religion have had mutually beneficial influences and will likely continue to do so.

---

5. Ferngren, *Medicine and Religion: A Historical Introduction.*

Science and theology aim to describe reality as a complete, self-consistent system. In practice, both endeavors use models to describe reality because the fabric of the universe contains aspects that are inherently difficult to understand. A classic example is the description of light as a wave and a particle, two mutually exclusive models that collectively describe light. In the same way theology also uses models, in this case to describe divine engagement. Models necessarily lead to a provisional understanding that are refined over time in a quest to most closely arrive at a true and accurate description of the world.

A provisional approach to science and religion is helpful because neither domain is completely objective. Michael Polyani successfully challenged the idea of an objective scientific method mechanically yielding truth in his book *Personal Knowledge* where he demonstrates the role of personal commitment in problem selection, the evaluation of experimental observations, theory development, and the influence of peer review in arriving at a scientific explanation. In practice, scientists may adopt, or refuse to adopt, new ideas for personal reasons rather than scientific merit. In the words of Max Planck "a new scientific truth does not triumph by convincing its opponents and making them see the light, but rather because its opponents eventually die, and a new generation grows up that is familiar with it."[6]

The discourse between science and religion exists in a continuum that has moved from independence to dialogue and is now focused on integration. Some people, particularly those who grew up with fundamentalist tendencies, tend to the independence end of the spectrum whereas others see less confrontation between science and religion. Part of the change, in line with Planck's assertion, may be because young people have an affinity for spirituality that fosters integration. Most theologians and scientists involved at the interface of the two domains pursue an integration approach because this view provides the most comprehensive explanation of reality.

DOES RELIGION MAKE ANY DIFFERENCE?

Assuming there is a religious dimension to life, and that the most significant locus of divine interaction with humanity is at the level of the individual, leads immediately to the question of whether divine action

---

6. Planck, *Scientific Autobiography and Other Papers*, 33–34.

is detectable in people's lives. Establishing good experimental protocols is challenging because of the difficulty in controlling the independent variables: an individual's religiosity, the link between belief and practice, and the numerous environmental influences capable of influencing the experimental outcome. Consequently, some studies do, while others do not, show a correlation between belief and well-being.

Behavioral scientists have sought to find evidence for an influence of religion in physical health. There has long been a correlation between religion and mental health, starting with Freud's assertion that religion is a neurosis based on people's wish that a being exists to care for and protect individuals. Review of the literature suggests a mixed, but overall helpful, effect of religion on physical health.[7] For example, some studies show that higher levels of religiosity correlate with lower rates of heart disease, stroke, kidney failure, and cancer mortality, as well as lower blood pressure. In one study, senior church attenders were found to have lower levels of interleukin 6—a chemical associated with disease, aging, and a decreased ability of the immune function to ward off illness—than the general population. There have been positive associations of religion and spirituality with well-being, marriage, meaning and purpose, and self-esteem, but there also exists a correlation of religion and spirituality with an increased tendency toward suicide, drug and alcohol abuse, criminality and depression. On balance, the influence of religion on mental health is usually judged to be positive.

One explanation for the positive influence of religion on physical and mental health is that the social network and support available through the association are helpful to people. Religious and spiritual communities provide opportunities for fellowship with community members in an environment where care for other people is expected. Church communities in particular stress the importance of making connections to other people through spiritually meaningful interactions that are focused on the divine. Efforts to increase religious commitments reinforce religious lifestyles that benefit physical and mental health. In general, the more conservative religious practitioners tend to have better physical and mental health.[8] Physiological factors have also been implicated in the link between religiosity and health. Practices such as forgiveness, hope, love, meditation, and contentment promote positive emotions that are linked

---

7. Koenig et al., *Handbook of Religion and Health*.
8. Ibid.

to good health. Religion seems to set in place a series of physical and mental practices that are beneficial to personal health while providing coping mechanisms that help address and overcome physical ailments. Religious people might claim that the benefit of religious practice is because people are made to be in relationship with God, and only when they are is their whole being healthy.

IS THE FORCE REALLY WITH US?

Typically religions view certain sages to have passed indispensable knowledge of God down to a chosen people. Christians believe that Jesus revealed God's essence not just by being his messenger, but by taking on human form, leading an exemplary life, and rescuing humanity from their spiritually broken nature. The revelation Jesus brings is that God wants each person to freely choose to engage in a spirit-directed life. The Bible essentially provides stories of divine relationships for individuals to learn from, which culminates in the story of God sending Jesus Christ as a bridge between God and humanity. The biblical description of God's human qualities, and having God come to earth as man, facilitates a personal relationship with God.

Science's method is focused on the observable world which necessarily excludes the personal influences and relationships that characterize the lived human experience. Great music is more than frequencies in the air; the Mona Lisa is more than an arrangement of oils and pigments; love and affection are more than a chemical release of hormones. The material aspects of life convey partial meaning that is complemented by the intimate moments of intense love and kindness that are central to religious belief.

If the goal of the cosmos is to evolve conscious beings capable of relationship with God then clues about God's presence are expected. The universe is not perfect as might be expected of an omnipotent creator but ironically the imperfections are often the places where God's presence is most readily detected. The lives of people suffering and experiencing loss can be enriched by God's presence in ways difficult to rationalize unless the world really is more than just a material universe.

The God of reconciliation described in the Bible is above all a God of love, love bestowed upon all through relationship. The insights of science can enable religious believers to know a God who acts at all levels of the physical universe, who is capable of creating a universe 13.7 billion years old and whose attention to detail extends to the smallest quantum particle. Perhaps most difficult to comprehend is the divine self-limitation that allows individuals to choose whether or not to reciprocate the interaction and engage in a personal relationship.

THEMES AND INFERENCES

Science is exceptionally successful in providing intricate details of the mechanism by which the universe evolved. Cosmology provides an explanation for the universe's unfolding from the Big Bang to create stars and planets. Inherent in cosmology is the extreme specificity of fundamental physical forces: gravity, electromagnetism, and the weak and strong nuclear forces. Accompanying this specificity are a series of unlikely events that are thought to be central in the creation of a hospitable earth; the chance impact of an enormous meteor hitting early earth which, on the rebound formed a moon essential to slowing the earth's rotation and creating tides—both important in evolution. Similar chance events seem to permeate the development of the earliest life forms and their progression to sentient complex individuals. Dinosaurs became extinct and were replaced by mammals, which ultimately evolved into modern humans. These are just a few examples of the chance events that are thought to have shaped evolutionary history to arrive at the world of today.

Linking together the sequence from Big Bang to humanity entails a remarkable series of events with a vanishingly small probability. The universe seems endowed with a weighted beneficence. Is this the result of providential divine activity written into the deep structure of the universe or a remarkable coincidence?

In the light of the scientific discoveries of the last five centuries, people in the western world have come to see themselves less as privileged individuals living at the center of the universe and more as primates located in a relatively uneventful part of the Milky Way, one of trillions of such galaxies in the observable universe. The shift occurred first in astronomy, then in biology, in physics in the last century, and now in neuroscience. The realization that humans are not privileged observers of the universe has influenced at least some religious thinkers to return theology to an emphasis on relationship between the individual and God rather than a physical presence at the center of the universe. Not that the centrality of a divine relationship was ever absent but rather that the place and significance of relationship between the individual and the divine—the focus of the biblical narrative—has tended to be obscured at times in the history of the church.

CONCLUSION

Science has been an incredible tool for advancing the frontiers of knowledge and understanding how the universe works. Science has provided insight into the human condition by breaking down each aspect of life into small isolatable units. A scientific explanation of the warm attraction between two people that blossoms into love may describe hormonal chemical interactions, but this is only a partial description of the human experience. Somehow, a greater understanding at the molecular level loses a key part of the relationships that are an essential part of being human.

Scientific and religious ways of knowing appear to be opposites. Scientific thinking focuses on knowing observable data while religious thinking involves revelation, tradition, and personal experience. Different kinds of information are internalized in different ways, which does not mean that there is a complete divide between the two realms. Like people suffering from depression, who look for both medical intervention and personal counselling, most people willingly accept a composite picture in their daily lives.

The human experience goes far beyond the measurable quantities that science can record or the physical difficulties of life that science and technology can overcome. Being human is fundamentally personal and subjective. People feel warmth from the sun, while physics measures the rays, people see friends and relatives while biology evaluates their gene propagation, people pray to and worship God for his hand in their lives while anthropology examines artifacts pointing toward cultural and religious beliefs.

Life persists in the midst of a struggle, but that very struggle can be both beautiful and elevated by grace. Recognizing nature's complexity, elegance, and beauty can be interpreted as an indirect recognition, and glorification of God, a recognition of divine thought inherent in the structure of the universe.

Is there a God? Many people believe that God exists or might exist, and choose to engage a relationship with him. Such a relationship is a freely chosen response to God. In this way, God is vulnerable because he cannot count on an individual's loyalty the way a person can rely on steadfast divine love.

Experiencing the world through an individual relationship with God offers the possibility of a more holistic understanding of reality than can be had from scientific observation and deduction alone. Incorporating a spiritual dimension into a materialistic perspective requires moving from testing to trusting. For those who have not begun on this path, the challenge is to run the experiment—to pray for God to reveal his presence in a way that is personal, relevant, and transformative. God's response will be the beginning of a relationship that enriches lives by revealing divine engagement now and in the life to come.

DISCUSSION QUESTIONS

1. Discussions of science and religion are often complicated because individuals have different conceptions of religion. Religion may be defined as the belief system of faith and worship, but philosopher-theologian John Cobb suggests that there is no common religious essence. Cobb believes religions are better understood as cultural movements with religious components. Reflecting on the influence of religious groups in the US and abroad, which definition do you think is more accurate?

2. Science, like religion, is hard to define. One workable definition provided by Fr. George Coyne is that science is the study of natural events by natural causes. The definition captures both biological and cultural aspects of science such as genetic evolution and the influence of culture on evolution that Richard Dawkins has coined memes. The definition limits science to the natural world while intellectually providing space for God to operate. Do you agree with this definition or should a comprehensive definition of science exclude divine action?

3. If you were God, how would you communicate God's love and desire for people to follow him to people living in diverse cultures over thousands of years?

4. Does the universe appear to be designed, and if so, what does this mean for the individual?

5. Is science morally neutral?

6. The current geological period is termed the Anthropocene period because of the significant human impact on modern history. For the first time in the history of the universe the future of life on earth now lies more in the hands of mankind than any other entity. How does humanity's scientific influence on earth, good and bad, relate intersect with the existence of a God constantly interacting with the world for the benefit of all people?

7. Assuming that God made the earth specifically for the human drama of life culminating in Christ's resurrection and ascension, is the existence of intelligent life in other parts of the universe reasonable?

8. The SETI project has been searching for intelligent life in the universe for over thirty years. If intelligent life were discovered in another galaxy, would this life form be expected to experience similar veiled evidence for God's existence?

Further reading for "Where Science and Religion Meet: Is there Personal Relevance?"

1 Alister McGrath, *The Science of God: An Introduction to Scientific Theology*. Grand Rapids: Eerdmans, 2004. Alister McGrath, who holds doctorates in molecular biophysics and historical and systematic theology, has written a comprehensive treaty developing scientific theology as a new philosophical mode of inquiry. The *Science of God* is a distillation of his three-volume treatise that provides a historical context to natural theology and then develops a scientific theology of nature and reality. McGrath uses his broad experience to provide a fundamental introduction for novitiates wanting to begin studying the interface of science and religion. As is typical of McGrath's work, the text provides a penetrating analysis of cultural and philosophical themes that is solid and well referenced.

2. John Polkinghorne, *Exploring Reality: The Intertwining of Science and Religion*. New Haven: Yale University Press, 2007. John Polkinghorne is one of the pre-eminent scholars writing at the interface of science and Christianity. This short book summarizes many of

his past ideas by surveying the insights of science and the insights of religion. Collectively this provides the foundation for discussing the ways the two interact practically in the context of belief.

3. Ian Barbour, *Religion and Science: Historical and Contemporary Issues.* New York: HarperCollins, 1997. Ian Barbour was one of the leading thinkers in science and religion with this revised book summarizing many of the ideas that helped reinvigorate the field. Of the four parts to the book, the second part explores different models relating science and religion.

4. Fraser Watts and Christopher Knight, eds., *God and the Scientist: Exploring the Work of John Polkinghorne.* Farnham, UK: Ashgate, 2012. The collected essays focus on the influence of John Polkinghorne's work in science and religion which provides a helpful overview of current issues at the forefront of the field.

5. Paul Davies, *The Mind of God: Science and the Search for Ultimate Meaning.* New York: Simon & Shuster, 1992. Paul Davies is unusual in being a mathematical physicist with a deep appreciation of the awe and beauty in the world begging for a metaphysical explanation but not personally believing in God. The essence of this book is to use scientific logic in answering the meaning of existence. Building on modern cosmology, Davies asks if the universe creates itself. He also explores where the laws of nature come from as well as beauty, elegance, and the permeation of mathematics through creation. True to being an open scientist this classic work looks from physics to mysticism for understanding the meaning of life.

6. Lyall Watson, *Dark Nature: A Natural History of Evil.* New York: HarperCollins, 1995. A fascinating book that tries to pin evil solely in nature rather than in the human heart. The naturalistic approach is one embraced by many scientists and Watson raises many knotty philosophical questions. A readable and thought provoking book.

7. Justin Barrett, *Why Would Anyone Believe in God?* Walnut Creek: Altamira, 2004. Barrett analyzes modern advances in cognitive science to understand where belief comes from. Materialist and evolutionary perspectives are included without judging the veracity of underlying assumptions, but rather focusing on explaining where beliefs in God come from.

8. Deborah B. Haarsma and Loren D. Haarsma, *Origins*. Grand Rapids: Faith Alive Christian Resources, 2007. Provides an excellent introduction from a reformed perspective. Different positions are presented in a non-judgmental manner showing advantages and challenges faced by each position.

9. Owen Gingerich, *God's Universe*. Cambridge: Harvard University Press, 2006. As a historian of science Gingerich brings a depth of knowledge and a wisdom of understanding to cosmology and the question of relevance. Chapter 1 is an excellent survey of the Copernican principle and the relevance to understanding. The style is engaging with sufficient background to understand the relevance without being bogged down in the details as illustrated in the chapter title: "Is Mediocrity a Good Idea?"

10. Peter Harrison, ed., *The Cambridge Companion to Science and Religion*. New York: Cambridge University Press, 2010. The collection of essays by leading philosophers, scientists, and theologians, explores contemporary issues of science and religion. The essay "Evolution and the Inevitability of Intelligent Life" by Simon Conway Morris explores the theme of biological evolution guided by natural law akin to that of gravity.

11. Lawrence Principe, *The Great Courses: Science and Religion*. Baltimore: The Teaching Company 2006, Audio. Philosophy and intellectual history professor Principe provides an accessible overview of the major points of interaction of science and religion from a historical perspective. Lecture two, *The Warfare Thesis*, provides a particularly good analysis of the roles played by the leading protagonists of the thesis, John Draper and Andrew Dickson White.

12. Michael Dodds, *Unlocking Divine Action: Contemporary Science and Thomas Aquinas*. Pittsboro NC: Catholic University of America Press, 2012. Dodds uses extensive quotations to summarize recent contributions to science and religion centering on divine action. Chapters 4 and 5 are particularly rich in understanding the challenges and approaches at the forefront of theological models that describe divine action in the world.

13. Ian Barbour, *Nature, Human Nature, and God*. Minneapolis: Augsburg Fortress, 2002. Barbour addresses some of the more difficult

recent topics at the interface of science and religion. Particularly insightful is the discussion of process theology.

9. Epilogue: Does Science Influence Personal Belief?

As a young boy living in rural New Zealand I can remember a sense of amazement in staring up at the glittering Milky Way. In the clear night sky, the stars appear to dance with a flickering intensity that gives the impression of them having a life of their own. From my Christian upbringing I knew that God had made the world but as I became older I saw the endless Milky Way as a possible witness for other intelligent life somehow outside God's boundaries.

Reading opened new viewpoints for explaining the world. Erich von Däniken's *Chariots of the Gods* provided compelling arguments for extraterrestrial life forms who were responsible for propelling humanity forward. Similar arguments were provided in the work of Arthur C. Clarke. I drank in these new ideas and championed them with anyone interested in finding the truth about humanity's place and purpose in the world.

I can vividly remember discussing these ideas with my Dad. He was an extremely intelligent person who graduated from high school but was denied further education because of difficult circumstances arising out of the Great Depression. I was particularly good at winning arguments for a world governed by mechanistic causes but I felt that my argument missed some essence that I could never quite identify. My Dad may have lost many of the arguments but he provided the answer in testifying that the purpose of life is only fulfilled by finding God. My Dad challenged me to test God's existence through prayer.

I prayed for God to perform acts that would undoubtedly confirm his existence. God never obliged. I reasoned that this neither proved nor disproved God's existence but merely confirmed my suspicion that God has created an orderly world where he prefers a subtle consistency in influencing the ephemeral world of people and relationships rather than through miraculous intervention in the physical world. My test for God's existence was caught in a paradox; if God did not provide a miracle then there would always be doubt and if he did I was not going to be convinced because my scientific mind would always invent a plausible materialistic explanation.

The real test for God's existence was to ask for personal direction in my life. With some trepidation, I asked God to help me decide whether to propose to the most wonderful young woman in all of New Zealand. Pam and I were married in my senior year of my BSc (Hons) degree and, after three decades, I can see that as being one of the most providential decisions I ever made.

As a skeptic, I knew that God needed to be tested some more. I had applied to over fifty PhD programs worldwide and prayed that God would lead my wife and me to the best place for us. Four and a half years later I graduated from the University of British Columbia, one of the best Canadian universities, where I was supported on a prestigious Killam Fellowship for four years. Providential again.

The time in British Columbia was probably the most formative time in my life. Unbeknownst to me I chose to work for a Christian professor, Edward Piers, who had a strong interest in trying to integrate his faith with his scientific understanding of the world. Ed was even more concerned in living a life he believed was worthy of being Christian, which was evident in his honesty, integrity, and true concern for others. For the first time my questions began to be illumined, not with glib answers but rather through a more holistic picture of the world that God created.

Was my choosing this particular professor another chance event or was this providence again? Chance upon chance upon chance or personal proof that God exists? Does God exist? Absolutely, prayer and blessing provide me with irrefutable proof. And that is again the problem. I look at these "chances" as providential. I now see the Milky Way as God's handiwork in providing for me and millions of others here on earth, but all these instances are my interpretation of events which others may simply interpret as luck or chance. The only way for anyone to find out is to run the experiment: to pray and ask God to reveal himself.

Bibliography

Alexander, Denis. *Creation or Evolution: Do We Have to Choose?* New York: Monarch, 2008.
Andrade, E. N. da C. *Rutherford and the Nature of the Atom.* New York: Doubleday, 1964.
Astin, Alexander W., et al. *Cultivating the Spirit: How College can Enhance Students' Inner Lives.* San Francisco: Jossey-Bass, 2010.
Augustine. *Confessions.* Translated by Edward Bouverie Pusey, 1909. Reprint. Project Gutenberg, 2002. Online: http://www.gutenberg.org/files/3296/3296-h/3296-h.htm.
Baglow, Christopher. *Faith, Science, and Reason: Theology on the Cutting Edge.* Chicago: Midwest Theological Forum, 2009.
Barbour, Ian. *Nature, Human Nature, and God.* Minneapolis: Augsburg Fortress, 2002.
———. *Religion and Science: Historical and Contemporary Issues.* New York: HarperCollins, 1997.
———. *When Science Meets Religion.* New York: HarperCollins, 2000.
Barr, Stephen. *Modern Physics and Ancient Faith.* South Bend, IN: University of Notre Dame Press, 2003.
Barrett, Justin. *Why Would Anyone Believe in God?* Walnut Creek, CA: Altamira, 2004.
Behe, Michael. *Darwin's Black Box.* 2nd ed. New York: Free, 2006.
Blomberg, Craig L. *The Historical Reliability of the Gospels.* Downer's Grove, IL: IVP, 1987.
Calaprice, Alice, and Trevor Lipscombe. *Albert Einstein: A Biography.* Westport, CT: Greenwood, 2007.
Chandrasekhar, S. "The General Theory of Relativity: 'Why It Is Probably the Most Beautiful of All Existing Theories.'" *Journal of Astrophysics and Astronomy* (1984) 3–11.
Cohen, Jon. *Almost Chimpanzee. Searching for What Makes us Human in Rainforests, Labs, Sanctuaries, and Zoos.* New York: Times, 2010.
Conway-Morris, Simon. *Life's Solution: Inevitable Humans in a Lonely Universe.* New York: Cambridge University Press, 2003.
Crick, Francis. *Astonishing Hypothesis: The Scientific Search for the Soul.* New York: Touchstone, 1995.
———. *Life Itself: Its Origin and Nature.* New York: Simon & Schuster, 1981.

Darwin, Charles. *The Origin of Species: 150th Anniversary Edition*. Tustin, CA: Mass Market Paperback, 2003.
Davies, Paul. *The 5th Miracle. The Search for the Origin and Meaning of Life*, New York: Touchstone, 1999.
———. *The Mind of God: Science and the Search for Ultimate Meaning*. New York: Simon & Shuster, 1992.
Dawkins, Richard. *The Blind Watchmaker*. New York: Norton, 1996.
de Duve, Christian. *Vital Dust: Life as a Cosmic Imperative*. New York: Basic, 1995.
de Waal, Frans B. W. *Good Natured: The Origins of Right and Wrong in Humans and Other Animals*. Harvard: Harvard University Press, 1996.
Denton, Michael. *Nature's Destiny: How the Laws of Biology Reveal Purpose in the Universe*. New York: Free, 1998.
Dodds, Michael. *Unlocking Divine Action: Contemporary Science and Thomas Aquinas*. Pittsboro, NC: Catholic University of America Press, 2012.
Dyson, Freeman. *Disturbing the Universe*. New York: Basic, 1979.
Einstein, Albert. "Religion and Science." *New York Times Magazine*, November 9, 1930. Online: http://einsteinandreligion.com/scienceandreligion2.html.
Ferguson, Kitty. *Tycho and Kepler: The Unlikely Partnership That Forever Changed Our Understanding of the Heavens*. New York: Bloomsbury, 2002
Ferngren, Gary B. *Medicine and Religion: A Historical Introduction*. Baltimore: Johns Hopkins University Press, 2014.
———. *Science and Religion: A Historical Introduction*. Baltimore: Johns Hopkins University Press, 2002.
Folsing, Albrecht. *Albert Einstein. A Biography*. New York: Viking, 2007.
Fox, Karen, and Aries Keck. *Einstein A to Z*. Hoboken, NJ: Wiley, 2004.
Galison, Peter, et al. *Einstein for the 21st Century: His Legacy in Science, Art, and Modern Culture*. Princeton: Princeton University Press, 2008.
Giberson, Karl, and Donald Yerxa. *Species of Origins: America's Search for a Creation Story*. Lanham, MD: Rowman and Littlefield, 2002.
Gingerich, Owen. *The Book Nobody Read: Chasing the Revolutions of Nicholas Copernicus*. New York: Penguin, 2005.
———. *The Eye of Heaven: Ptolemy, Copernicus, Kepler*. New York: American Institute of Physics, 1993.
———. *God's Universe*. Cambridge: Harvard University Press, 2006.
Gladwell, Malcolm. *The Tipping Point: How Little Things Can Make a Big Difference*. New York: Little, Brown, and Company, 2000
Grant, Edward. *Science and Religion, 400 BC to A.D. 1550*. Baltimore: The Johns Hopkins University Press, 2006.
Green, Joel B., and Stuart L. Palmer. *In Search of the Soul: Four Views of the Mind-Body Problem*. Downer's Grove, IL: IVP, 2005.
Haarsma, Deborah B., and Loren D. Haarsma. *Origins*. Grand Rapids: Faith Alive Christian Resources, 2007.
Hayes, Cathy. *The Ape in Our House*. New York: Harper, 1951.
Hannam, James. *The Genesis of Science: How the Christian Middle Ages Launched the Scientific Revolution*. Washington, DC: Regnery, 2011.
Harrison, Peter, ed. *The Cambridge Companion to Science and Religion*. New York: Cambridge University Press, 2010.

Hodge, Jonathan, and Gregory Radlick, eds. *The Cambridge Companion to Darwin*. 2nd ed. New York: Cambridge University Press, 2009.

Hoffmann, Banesh. *Albert Einstein: Creator and Rebel*. New York: Viking Adult, 1972.

Holder, Rodney. *Big Bang, Big God: A Universe Designed for Life?* Oxford: Lion Hudson, 2013.

———. *God, the Multiverse, and Everything: Modern Cosmology and the Argument from Design*. Farnham, UK: Ashgate, 2004.

Holton, Gerald. *The Scientific Imagination: Case Studies*. Cambridge: Cambridge University Press, 1978.

Hoyle, Fred. "The Universe: Past and Present Reflections." *Engineering & Science* (1981) 8–12.

Hummel, Charles. *The Galileo Connection*. Downers Grove, IL: IVP, 1986.

Isaacson, Walter. *Einstein: His Life and Universe*. New York: Simon and Shuster, 2007.

Jaki, Stanley. *The Road of Science and the Ways to God*. Chicago: University of Chicago Press, 1980.

Jeeves, Malcolm. *From Cells to Souls—And Beyond: Changing Portraits of Human Nature*. Grand Rapids: Eerdmans, 2004.

Jeeves, Malcolm, and R. J. Berry. *Science, Life, and Christian Belief: A Survey of Contemporary Issues*. Grand Rapids: Baker, 1999.

Jeeves, Malcolm, and Warren Brown. *Neuroscience, Psychology, and Religion: Illusions, Delusions, and Realities about Human Nature*. Conshohocken, PA: Templeton, 2009.

Jensen, Alexander S. *Divine Providence and Human Agency: Trinity, Creation and Freedom*. Farnham, UK: Ashgate, 2014.

John Paul II, Pope. *Letter of His Holiness John Paul II to Reverend George V. Coyne*. Rome: Libreria Editrice Vaticana, 1988. Reprint, http://w2.vatican.va/content/john-paul-ii/en/letters/1988/documents/hf_jp-ii_let_19880601_padre-coyne.html.

———. "Truth Cannot Contradict Truth." *Address to the Pontifical Academy of Sciences*. 1996. Online: http://www.beliefnet.com/News/Science-Religion/2000/03/Truth-Cannot-Contradict-Truth.aspx#.

Johnson, Phillip. *Darwin on Trial*. Washington, DC: Regnery, 1991.

Kahneman, Daniel. *Thinking Fast and Slow*. Reprint. New York: Farrar, Straus and Giroux, 2011.

Kellog, W. N., and L. A. Kellog. *The Ape and the Child*. New York: Hafner, 1967.

Knowles, Elizabeth, and Angela Partington, eds. *The Oxford Dictionary of Quotations*. New York: Oxford University Press, 1999.

Koenig, Harold G., et al. *Handbook of Religion and Health*. New York: Oxford University Press, 2001.

Kozhamthadam, Job. *The Discovery of Kepler's Laws: The Interaction of Science, Philosophy, and Religion*. South Bend, IN: University of Notre Dame Press, 1994.

Larson, Edward J. *Evolution's Workshop: God and Science on the Galapagos Islands*. New York: Basic, 2001.

Luisi, Pier Luigi. *The Emergence of Life from Chemical Origins to Synthetic Biology*. New York: Cambridge University Press, 2006.

Machamer, Peter. *The Cambridge Companion to Galileo*. New York: Cambridge University Press, 1999.

McGrath, Alister. *The Science of God: An Introduction to Scientific Theology.* Grand Rapids: Eerdmans, 2004.

Miller, Kenneth. *Finding Darwin's God: A Scientist's Search for Common Ground between God and Evolution.* New York: Harper Perennial, 1999.

Moreland, James P., and Scott B. Rae. *Body and Soul: Human Nature and the Crisis in Ethics.* Downer's Grove, IL: IVP, 2000.

Morison, Frank. *Who Moved the Stone?* Grand Rapids: Zondervan, 1987.

Moss, Jean Dietz. *Novelties in the Heavens: Rhetoric and Science in the Copernican Controversy.* Chicago: University of Chicago Press, 1993.

Naess, Atle. *Galileo Galilei—When the World Stood Still.* Heidelberg: Springer, 2005.

Newton, Isaac. *The Principia: Mathematical Principles of Natural Philosophy.* Reprint. New York: Snowball, 2010.

Nichols, Terence. *The Sacred Cosmos.* Grand Rapid: Brazos, 2003.

Olson, Richard. *Science and Religion. 1450–1900: From Copernicus to Darwin.* Baltimore: Johns Hopkins University Press, 2004.

———. *Science Deified & Science Defied: The Historical Significance of Science in Western Culture.* Berkeley: University of California Press, 1995.

Overbye, Dennis. "Physicists Find Elusive Particle Seen as Key to Universe." *New York Times,* July 5, 2012, A1.

Overman, Dean. *A Case Against Accident and Self-Organization.* Washington, DC: Rowman and Littlefield, 1997.

Paley, William. *The Works of William Paley.* London: Davison, 1830.

Planck, Max. *Scientific Autobiography and Other Papers.* Translated by F. Gaynor. New York: Philosophical Library, 1949.

Phipps, William. *Darwin's Religious Odyssey.* Harrisburg, PA: Trinity, 2002.

Plantinga, Alvin. *God, Freedom, and Evil.* Grand Rapids: Eerdmans, 1989.

Polkinghorne, John. *Exploring Reality. The Intertwining of Science and Religion.* New Haven: Yale University Press, 2007.

———. *Science and Providence: God's Interaction with the World.* Conshohocken, PA: Templeton Foundation, 2011.

———. "So Finely Tuned a Universe of Atoms, Stars, and Quanta, and God." *Commonweal,* August 16, 1996, 16.

———. *Theology in the Context of Science.* New Haven: Yale University Press, 2009.

Principe, Lawrence M. *The Great Courses: Science and Religion, Philosophy and Intellectual History.* The Teaching Company, 2006. Audio.

Rana, Fazale, and Hugh Ross. *Origins of Life: Biblical and Evolutionary Models Face Off.* Colorado Springs: NavPress, 2004.

Ridley, Matt. *The Rational Optimist: How Prosperity Evolves.* New York: HarperCollins, 2010.

Ross, Hugh. *Hidden Treasures in the Book of Job: How the Oldest Book in the Bible Answers Today's Scientific Questions.* Grand Rapids: Baker, 2011.

Ratzsch, Del. *Nature, Design, and Science: The Status of Design in Natural Science.* New York: SUNY, 2001.

Sagan, Carl. *Cosmos.* New York: Random House, 2002.

———. *Cosmos: A Personal Voyage.* Cosmos Studios, DVD Release Date: October 22, 2002.

Seybold, Kevin. *Explorations in Neuroscience, Psychology, and Religion,* Farnham, UK: Ashgate, 2007.

Southgate, Christopher, ed. *God, Humanity, and the Cosmos: A Textbook in Science and Religion.* New York: Trinity, 1999.
Spradley, Joseph. *Visions That Shaped the Universe: A History of Scientific Ideas about the Universe.* Dubuque, IA: Brown, 1995.
Stockwood, Mervyn, ed. *Religion and the Scientists.* London: SCM, 1959.
Stokes, Mitch. *Galileo.* Nashville: Thomas Nelson, 2010.
———. *Isaac Newton.* Nashville: Thomas Nelson, 2010.
Stoller, Steve. *The Symphony of Creation: Science and Faith in Harmony.* Phoenix: ACW, 2002.
Strassman, Rick. *DMT: The Spirit Molecule: A Doctor's Revolutionary Research into the Biology of Near-Death and Mystical Experiences.* South Paris, ME: Park Street, 2001.
Stringer, Chris. *The Origin of Our Species.* London: Penguin, 2011.
Stump, Eleonore, and Norman Kretzmann, eds. *The Cambridge Companion to Augustine.* New York: Cambridge University Press, 2001.
Swinburne, Richard. *Providence and the Problem of Evil.* Oxford: Clarendon, 1998.
Tattersall, Ian. *Becoming Human: Evolution and Human Uniqueness.* San Diego: Harcourt Brace, 1998.
———. *Paleontology: A Brief History of Life.* Conshohocken, PA: Templeton Foundation, 2010.
Thomson, Keith Stewart. *Private Doubt, Public Dilemma: Religion and Science since Jefferson and Darwin.* New Haven: Yale University Press, 2015.
Thuan, Trinh X. *Chaos and Harmony: Perspectives on Scientific Revolutions of the Twentieth Century.* Conshohocken, PA: Templeton Foundation, 2006.
Temerlin, Maurice K. *Lucy: Growing Up Human: A Chimpanzee Daughter in a Psychotherapist's Family.* Mountain View, CA: Science & Behavior Books, 1976.
Vatican. "Catechism of the Catholic Church." Online: http://www.vatican.va/archive/ccc_css/archive/catechism/p1s1c3a1.htm.
Voelkel, James. *Johannes Kepler and the New Astronomy.* New York: Oxford University Press, 1999.
Ward, Keith. *God, Chance, and Necessity.* London: Oneworld, 1996.
Ward, Peter D., and Donald Brownlee. *Rare Earth: Why Complex Life is Uncommon in the Universe.* New York: Springer, 2007.
Watts, Fraser, and Christopher Knight, eds. *God and the Scientist: Exploring the Work of John Polkinghorne.* Farnham, UK: Ashgate, 2012.
Watson, Lyall. *Dark Nature. A Natural History of Evil.* New York: HarperCollins, 1995.
Willard, Dallas. *The Divine Conspiracy: Rediscovering Our Hidden Life in God.* New York: Harper, 1998.
Westfall, Richard S. "Isaac Newton." In *Science and Religion: A Historical Introduction*, edited by Gary B. Ferngren, 153–62. Baltimore: Johns Hopkins University Press, 2002.
Wright, N. T. *The Challenge of Jesus: Rediscovering Who Jesus Was & Is.* Downer's Grove, IL: IVP, 1999.
Yancey, Philip. *The Jesus I Never Knew.* Grand Rapids: Zondervan, 1995.
Young, Christian C., and Mark A. Largent. *Evolution and Creationism: A Documentary and Reference Guide.* Westport, CT: ABC-CLIO, 2007.

Subject Index

1st law of motion, 133
2nd law of motion, 134
Adam, 85–86, 87, 90, 184
altruism, 81–82
anthropic principle, 10–11
Arabic science, 121–22
Aristotle, 119–20, 137–38
artificial intelligence, 167–68
Augustine, 8, 87, 103, 124
axioms in science, 32–34

Babylonian science, 117
Big Bang, 1–16
    Bible, 2
    time, 8
    chance or design, 9
boundary conditions, 4
Brahe, Tyco, 131–32
brain, 57, 68, 71, 78, 84, 165–75

causal joint, 104–7, 188
cells, 38–42, 46–47
chance, 9, 59, 203
chaos theory, 105–6
child versus chimp, 75–76
Chinese room, 166
coincidence, 2, 10, 102, 187, 203
complexity, 14–15, 35–36, 41–42, 53, 55, 56
computer brain, 166–69
conflict thesis *see* warfare model,
consciousness, 70, 173–76, 182

Copernicus, Nicolas, 126–29
cosmology, 6, 9, 139
creation, 21–25, 51–52, 56–59, 124
    time, 8
creationism, 14

Darwin, Charles, 50, 56, 64, 149–53
death, 50–53, 82, 108–10, 173, 181–84
Deep Blue, 167
descent with modification, 14, 152, 159
design, 9–11, 14–15, 55–56, 135, 152
determinacy, 59–61
Dialogue Concerning the Two Chief World Systems, 141
dinosaurs, 50–51
divine activity, 104–7
Draper, Jon William, 196
drugs, 176–78, 186
Duchess Christina, 139–40

Egyptian science, 116–17
Einstein, Albert, 33, 153–58
elliptical motion, 133
emergence, 171–73
evil, 51–53, 86, 104
    natural, 80–83
    moral, 86–89
evolution, 20–42, 46–63, 67–90
ex nihilo, 124

fine tuning, 4–7
fossil, 14, 27, 47, 49
free will, 51–52, 171–73, 178–81

Gage, Phineas, 179
Galilei, Galileo, 136–44
Garden of Eden, 51, 86, 90
Genesis, 3, 21–25, 67, 85
   1:1, 2
   1:1–31, 21–23
   2:20, 51
   1:24–25, 57
   3:1, 86
God-of-the-gaps, 43
God helmet, 187
God particle, 5
Gödel, Kurt, 168
Grand Unifying Theory (GUT), 10–11
Greek science, 118–21

habitable zone, 12–13
Heisenberg's uncertainty principle, 60, 105
Higgs boson, 6
hominids, 67–90
Hoyle, Fred, 2, 6
Hubble, Edwin, 2–3
human development, 78–80
humans, 77–78, 184

Indian science, 123
information, 34–37, 41, 47, 105, 174, 185
integration model, 199–200
intelligent design, 55–56, 197
irreducible complexity, 55
Islam, 122–23

Jesus, 95–112, 184–85

Kepler, Johannes, 129–36

language, 68–69, 74–76
Lemaître, George, 3

life, 11–13, 20–43, 46–50, 57–58, 60, 97–98, 152, 173

machine intelligence, 167
Maric, Mileva, 155
mathematics, 4, 6, 10, 15, 33
microwave radiation, 2
Miller-Urey experiment, 29
mind, 76, 79, 97, 110, 169–71, 173–78
miracles, 101–4, 107, 111
mirror neurons, 84
modern man, 73–74
mood altering drugs, 78, 177–78
morality, 83–84, 175, 179
MRI, 84, 180
mutation, 41, 52–54
mystical experiences, 185–89

natural evil, 52, 80–83
natural selection, 53–54
natural theology, 14–15
Neanderthals, 72–73, 78
near death experience, 181–82
Newton, Isaac, 144–49
non-overlapping magisterial approach, 198

Origin of Species, 56
Osiander, Andreas, 128

pain, 51–52, 82–83, 88
Paley, William, 14
parable, 97–98
prayer, 99–101, 104–6, 188–89, 210
prebiotic evolution, 28–30
prefrontal cortex, 178–80
primates, 67–70, 80, 84
primordial Earth, 29
Principia, 146–47
process theology, 104
Providence, 56, 101–2, 130, 149
Ptolemy, 120–21

radioactive dating, 26

relationships, 82, 90, 98, 135, 156, 171, 183–84, 202
religious experiences, 165, 186, 188
replication, 37–38
resurrection, 108–10, 184, 185
Roman science, 121–22

Scheiner, Father Christopher, 139
Scopes trial, 197
separate spheres, 198
sin, 80, 84, 87–88, 110, 184
sociobiology, 81
soul, 183–85

spiritual experiences, 185–89
star birth, 5, 7–9
stromatolites, 47
suffering, 50–53, 63, 82, 88

The Force, 202–3
time, 8–9, 24–25, 117
Trinity, 135, 147

warfare model, 195–97
water, 27–28, 29
White, Andrew Dickson, 196–97

www.ingramcontent.com/pod-product-compliance
Lightning Source LLC
Chambersburg PA
CBHW062020220426

43662CB00010B/1406